Probability and Its Applications

R.K. Getoor

Excessive Measures

Birkhäuser
Boston · Basel · Berlin

R.K. Getoor
Department of Mathematics
University of California, San Diego
La Jolla, CA 92093, USA

Library of Congress Cataloging-in-Publication Data
Getoor, R.K. (Ronald Kay), 1929–
Excessive measures/R.K. Getoor.
p. cm.—(Probability and its applications)
Includes bibliographical references.
ISBN 0-8176-3492-4 (alk. paper)
1. Excessive measures (Mathematics) 2. Markov processes.
3. Measure theory. 4. Potential theory (Mathematics) I. Title.
II. Series.
QA274.7.G47 1990 89-18654
519.2′33—dc20

Printed on acid-free paper.

ISBN 0-8176-3492-4
ISBN 3-7643-3492-4

Camera-ready copy supplied by author using $\TeX$.
Printed and bound by Edwards Brothers, Inc., Ann Arbor, Michigan.
Printed in the U.S.A.

9 8 7 6 5 4 3 2 1

Preface

The study of the cone of excessive measures associated with a Markov process goes back to Hunt's fundamental memoir [**H57**]. However until quite recently it received much less attention than the cone of excessive functions. The fact that an excessive function can be composed with the underlying Markov process to give a supermartingale, subject to secondary finiteness hypotheses, is crucial in the study of excessive functions. The lack of an analogous construct for excessive measures seemed to make them much less tractable to a probabilistic analysis. This point of view changed radically with the appearance of the pioneering paper by Fitzsimmons and Maisonneuve [**FM86**] who showed that a certain stationary process associated with an excessive measure could be used to study excessive measures probabilistically. These stationary processes or measures had been constructed by Kuznetsov [**Ku74**] extending earlier work of Dynkin. It is now common to call them Kuznetsov measures. Following the Fitzsimmons-Maisonneuve paper there was renewed interest and remarkable progress in the study of excessive measures. The purpose of this monograph is to organize under one cover and prove under standard hypotheses many of these recent results in the theory of excessive measures.

The two basic tools in this recent development are Kuznetsov measures mentioned above and the energy functional. The energy functional has a long history that may be traced back to Hunt, but its systematic use in the study of excessive measures seems to be more recent. However, see [**CL75**] for its definition and use in an abstract setting. Also it was used for other purposes by Meyer in [**Me68**] and [**Me73**]. A third ingredient in this development is the use of two Riesz type decompositions of an excessive measure: the first into dissipative and conservative parts is due to Dynkin [**Dy80**]—see also Blumenthal [**B1**]; the second into a potential and a harmonic part is due originally to Getoor and Glover [**GG84**]. Both of these decompositions, as well as the more elementary decomposition into purely excessive and invariant parts, were given probabilistic interpretations in terms of Kuznetsov measures in [**FM86**].

Using these tools one can construct a potential theory for excessive measures that in many respects is closer to classical potential theory than the potential theory of excessive functions. In classical potential theory or, more generally, under strong duality assumptions as in Chapter VI of [**BG**], there is an isomorphism between excessive measures and (a class of) coexcessive functions. It turns out that many of the fine results about excessive functions under these hypotheses, when interpreted as theorems about coexcessive measures, have a natural extension to a general Markov process, even though the corresponding results for excessive functions do not generalize completely. Thus the natural generalization of certain classical results only appears in the potential theory of excessive measures. Of course, there exist generalizations of the classical theory to abstract cones that include both excessive measures and functions. However, our emphasis here is on the underlying probabilistic meaning of the potential theory.

One other important benefit of this approach is that it requires no a priori transience hypothesis: the transience assumptions being subsumed under the conservative-dissipative dichotomy. It is remarkable that often the results are the same in the two cases, although the proofs may be quite different.

In the first five sections we develop the theory of excessive measures about as far as we can without using Kuznetsov measures. In sections 6 and 7 we introduce Kuznetsov measures and use them to study excessive measures and their potential theory. These first seven sections contain the basic potential theory of excessive measures. Sections 8, 10 and 11 contain other important applications of the energy functional and Kuznetsov measures, but the role of excessive measures is somewhat secondary. Section 9 on flows and Palm measures is perhaps tangential to the main development, but is important for a better understanding of sections 8 and 10. Appendix A contains an expanded proof of Meyer's perfection theory for multiplicative functionals [**Me74**]. Although one could avoid it—a cost, of course—I have decided to include it because of its importance. It is often quoted and is deserving of an expanded proof.

My guiding principle during the writing was to give complete proofs of all results that are not available in the standard reference books listed at the beginning of the bibliography. Like most principles, it is easier to formulate in the abstract than follow in the particular. One consequence of this is that I refer to these standard books for needed facts whenever possible, rather than to the original papers in which they appeared. (As of this writing only a preliminary version of Chapters XVII and XVIII of the final part of the monumental treatise [**DM**] by Dellacherie and Meyer is available to me, and so references to these two chapters may not be completely accurate when the definitive version appears.)

It is a pleasure to acknowledge a few of my debts. First and most of all I must thank Pat Fitzsimmons. Even the casual reader will notice the extent to which his ideas pervade this work. But I owe him much more. I had the privilege of consulting with him on an almost daily basis during the writing of this volume. Time and time again he set me straight and pointed the way when I was stuck or confused. It seems unlikely that this work would have been completed without his help. I would especially like to thank him for his contributions to Appendix A and to acknowledge that the concept of a "good partition" used there is due to him. Last but not least he read the entire manuscript and made numerous suggestions for improving the exposition. Jutta Steffens supplied the key step in the proof of Proposition 4.17. This enabled me to simplify considerably the original proof of the important Theorem 7.9 which gives the probabilistic meaning of Hunt's balayage operation on excessive measures. Neola Crimmins displayed her customary superb skill as well as unlimited patience in transforming my handwritten scrawl into the beautiful TeX format. Finally I received support from the National Science Foundation under NSF Grant DMS87-21347 during part of the writing.

La Jolla, California R. K. Getoor
September, 1989

Contents

Preface v

1. Notation and Preliminaries 1
2. Excessive Measures 6
3. The Energy Functional 16
4. Balayage of Excessive Measures 22
5. Potential Theory of Excessive Measures 35
6. Kuznetsov Measures 50
7. Kuznetsov Measures II 68
8. Homogeneous Random Measures 80
9. Flows and Palm Measures 95
10. Palm Measures and Capacity 116
11. Exit Systems and Applications 136

Appendix A 151

Appendix B 174

Bibliography 180

Notation Index 185

Subject Index 187

1. Notation and Preliminaries

We shall assume once and for all that

$$X = (\Omega, \mathcal{F}, \mathcal{F}_t, X_t, \theta_t, P^x)$$

is a right Markov process as defined in §8 of [**S**] with state space $(E, \mathcal{E})$, semigroup (P_t), and resolvent (U^q). To be explicit E is a separable Radon space and $\mathcal{E}$ is the Borel σ-algebra of E. A cemetery point Δ is adjoined to E as an isolated point and $E_\Delta := E \cup \{\Delta\}$, $\mathcal{E}_\Delta := \sigma(\mathcal{E} \cup \{\Delta\})$. (The symbol "$:=$" should be read as "is defined to be".) We suppose that [**S**, (20.5)] holds; that is, $X_t(\omega) = \Delta$ implies that $X_s(\omega) = \Delta$ for all $s \geq t$ and that there is a point $[\Delta]$ in Ω (the dead path) with $X_t([\Delta]) = \Delta$ for all $t \geq 0$. Of course, $\zeta = \inf\{t: X_t = \Delta\}$ is the lifetime of X. The filtration $(\mathcal{F}, \mathcal{F}_t)$ is the augmented natural filtration of X, [**S**, (3.3)]. We shall always use $\mathcal{E}$ to denote the Borel σ-algebra of E in the original topology of E. Beginning in §20, Sharpe uses $\mathcal{E}$ to denote the Borel σ-algebra of E in the Ray topology. We shall *not* use this convention. We shall write $\mathcal{E}^r$ for the σ-algebra of Ray Borel sets. These assumptions on X are weaker than those in [**G**] or [**DM**, XVI-4]. Beginning in §6 we shall make an additional assumption on X. (See (6.2)). To avoid trivialities *we assume throughout this monograph that* $X_\infty(\omega) = \Delta$, $\theta_0\omega = \omega$, *and* $\theta_\infty\omega = [\Delta]$ *for all* $\omega \in \Omega$.

Our notation for the objects associated with X is the standard notation (with few exceptions) that may be found in the familiar reference books [**BG**], [**DM**], [**G**], and [**S**]. For the convenience of the reader we shall recall the basic ones as well as some of their familiar properties. We refer the reader to the above references for proofs.

If $(H, \mathcal{H})$ is a measurable space, $\mathcal{H}^*$ denotes the σ-algebra of universally measurable sets over $(H, \mathcal{H})$. If $\mathcal{C}$ is any collection of extended real valued functions on H, then $p\mathcal{C}$ (resp. $b\mathcal{C}$) denotes those $f \in \mathcal{C}$ which are positive (resp. bounded). In particular $p\mathcal{H}$ (resp. $b\mathcal{H}$) denotes the collection of positive (resp. bounded) $\mathcal{H}$ measurable functions on H.

S^q or $S^q(X)$ denotes the cone of q-excessive functions of X, $q \geq 0$. As usual we write $S = S^0$. Perhaps the most useful σ-algebra on E is $\mathcal{E}^e := \sigma\left(\bigcup_{q\geq 0} S^q\right)$. It is immediate from the resolvent equation that $\mathcal{E}^e = \sigma(S^q)$ for any $q > 0$. Also P_t and U^q map $\mathcal{E}^e$ into itself. The potential kernel of X, $U := U^0$ is *proper* provided there exists $f \in \mathcal{E}^*$ with $f > 0$ and $Uf < \infty$. Then there exists $g \in \mathcal{E}^e$ with $g > 0$ and $Ug \leq 1$. See [**DM**, XII-8] or [**G80**]. In fact one may even suppose that g is finely continuous. If U is proper and $u \in S$, then there exists a sequence $(f_n) \subset pb\mathcal{E}^e$ with $Uf_n \uparrow u$ [**DM**, XII-17]. The process X is *transient* provided U is proper. A function $f \in p\mathcal{E}^*$ is *supermedian* provided $P_t f \leq f$ for each $t \geq 0$. Then $\hat{f} := \uparrow \lim_{t\downarrow 0} P_t f$ is excessive and $\hat{f}$ is called the *excessive regularization* of f. We write $P_t^q = e^{-qt}P_t$ for $q > 0$. This is the semigroup of X^q, the q-subproces of X. Clearly $S^q(X) = S(X^q)$. Since X^q is a right process all of the above considerations may be applied to X^q. In particular, X^q is transient when $q > 0$. We let $\mathcal{F}_t^e$ (resp. $\mathcal{F}^e$) denote the σ-algebra generated by $f \circ X_s$ with $f \in \mathcal{E}^e$ and $s \leq t$ (resp. $s < \infty$).

A σ-finite measure ξ on $(E, \mathcal{E})$ is *q-excessive* provided $\xi P_t^q \leq \xi$ for each $t \geq 0$. From now on we shall just say ξ is a measure on E when we mean a measure on $(E, \mathcal{E})$. Of course, any measure ξ on E has a unique extension to $\mathcal{E}^*$ which we again denote by ξ. Let Exc^q or $\mathrm{Exc}^q(X)$ denote the class of q-excessive measures. We drop q from the notation when it has the value zero. Thus Exc denotes the class of excessive measures. Obviously $\mathrm{Exc}^q(X) = \mathrm{Exc}(X^q)$. If $\xi \in \mathrm{Exc}^q$, then

$\xi P_t^q \uparrow \xi$ as $t \downarrow 0$ [**DM**, XII-37b]. Suppose $\xi \in \mathrm{Exc}$ and U is proper. Then there exists a sequence of finite measures (μ_n) on E with $\mu_n U \uparrow \xi$ [**DM**, XII-38]. Clearly if $\xi, \eta \in \mathrm{Exc}$, then $\xi \wedge \eta \in \mathrm{Exc}$ and if (ξ_n) is an increasing sequence of excessive measures, then $\xi := \uparrow \lim \xi_n \in \mathrm{Exc}$ provided ξ is σ-finite.

If $B \in \mathcal{E}^e$, then $T_B := \inf\{t > 0: X_t \in B\}$ where the infimum of the empty set is $+\infty$ is an $(\mathcal{F}_t)$ stopping time. If $B \in \mathcal{E}$, then T_B is even an $(\mathcal{F}_{t+}^*)$ stopping time; that is $\{T_B < t\} \in \mathcal{F}_t^*$ for each $t > 0$. Here $\mathcal{F}_t^*$ is the universal completion of $\mathcal{F}_t^0 := \sigma(X_s: s \leq t)$ and $\mathcal{F}_t^* \subset \mathcal{F}_t$ for each t. T_B is the *hitting time* of B. Define for $f \in p\mathcal{E}^*$

$$(1.1)\quad P_B^q f(x) = P^x[e^{-qT_B} f \circ X_{T_B}] := \int e^{-qT_B} f \circ X_{T_B}\, dP^x .$$

where, by convention, $e^{-0T_B} = 1_{\{T_B < \infty\}}$. We also use the convention that any function f on E is extended to E_Δ by $f(\Delta) = 0$. Thus in (1.1) the integration in ω is only over the set $\{T_B < \zeta\}$. Clearly P_B^q is a kernel on $(E, \mathcal{E}^*)$. If μ is a finite measure on E and $B \in \mathcal{E}^e$, then there exists an increasing sequence (K_n) of compact subsets of B with $T_{K_n} \downarrow T_B$ a.s. P^μ [**G**, (12.15)]. If $B \in \mathcal{E}^e$, B^r denotes the set of regular points for B, $B^r := \{x: P^x(T_B = 0) = 1\}$ and $B^r \in \mathcal{E}^e$. It is important to note that B^r and P_B^q depend only on the semigroup (P_t) and not on the particular realization, X, of (P_t) as a right process. See §19 of [**S**]; especially (19.7) in which nearly optional should be replaced by $\mathcal{E}^e$ measurable. (The proof of (19.7ii) as stated in [**S**] is incomplete.)

The definition of the fine topology in [**S**] should be modified slightly. The definition (10.7) in [**S**] should be changed to read as follows: A subset G of E is *finely open* provided for each $x \in G$ there exists $B \in \mathcal{E}^e$ such that $x \in B \subset G$ and $x \notin (B^c)^r$. (Here $B^c := E \backslash B$ and our B corresponds to B^c in [**S**, (10.7)]. The critical difference is that we require $B \in \mathcal{E}^e$ rather than nearly optional.) By the zero-one law $x \notin (B^c)^r$ if and only if $P^x(T_{B^c} > 0) = 1$. The fine topology on E is the collection of finely open sets. With this definition the fine topology depends only on the semigroup (P_t) and not on the particular realization, X, of (P_t) as a right process. Consequently in proving results about the fine topology one may suppose X satisfies the conditions in §20 of [**S**] without loss of generality. With this modification in the definition of the fine topology, the proof of Theorem 49.9 in [**S**] is valid. It states that the fine topology is generated by $\bigcup_{q\geq 0} S^q$, or just by S^q for any $q > 0$. The proof of the following proposition is a straightforward adaptation of the proof of [**BG**, II-(4.6)].

(1.2) **Proposition.** *Suppose U is proper. Then the fine topology is generated by S.*

Proof. As observed earlier we may suppose X satisfies the conditions in §20 of [**S**] in the proof. Let $\mathcal{T}$ be the topology generated by S and let $\mathcal{O}(f)$ be the fine topology. Then it suffices to show $\mathcal{O}(f) \subset \mathcal{T}$. The proof of [**S**, (49.9)] shows that sets of the form $\{\psi_G < 1\}$ where $\psi_G := P^{\cdot}(e^{-T_G})$ as G ranges over all open sets in the Ray topology of E form a base for $\mathcal{O}(f)$. Fix such a G and let $\psi = \psi_G$. Since U is proper, there exists $0 < h \leq 1$ with $h \in \mathcal{E}^e$ and $Uh \leq 1$. Let $f_n = (nh) \wedge 1$. Then $f_n > 0$, $f_n \uparrow 1$, and $Uf_n \leq n$. Hence

$$\varphi_n := Uf_n - P_G U f_n = P^{\cdot} \int_0^{T_G} f_n \circ X_t \, dt \uparrow P^{\cdot}(T_G \wedge \zeta).$$

Consequently $\varphi := P^{\cdot}(T_G \wedge \zeta)$ is $\mathcal{T}$-l.s.c. If $\varphi(x) > 0$, then $P^x(T_G > 0) = 1$ and so $\psi(x) < 1$. But $P^x(T_G > 0) = 1$ if $\psi(x) < 1$ and thus $\varphi(x) > 0$. Therefore $\{\psi < 1\} = \{\varphi > 0\} \in \mathcal{T}$. Hence $\mathcal{O}(f) \subset \mathcal{T}$. ∎

We close this section with some additional notational conventions which we shall use without special mention in the sequel. Let $(M, \mathcal{M})$ and $(N, \mathcal{N})$ be measurable spaces. We write $f \in \mathcal{M}|\mathcal{N}$ to indicate that f is a measurable mapping from M to N; that is, $f: M \to N$ and $f^{-1}(\mathcal{N}) \subset \mathcal{M}$. If B is any subset of M, $\mathcal{M}|_B$ denotes the trace of $\mathcal{M}$ on B. If f is a numerical function on M, $\|f\| := \sup\{|f(x)| : x \in M\}$.

If μ is a measure on $(M, \mathcal{M})$ and $f \in p\mathcal{M}^*$ we write $f\mu$ or $f \cdot \mu$ for the measure $B \to \int_B f\,d\mu$ defined on $(M, \mathcal{M})$. We also write $\mu(f)$ or $\langle \mu, f \rangle$ for $\int f\,d\mu$, and sometimes just μf. Thus $\mu U f = \mu(Uf) = \mu U(f)$. All named functions on E — the state space of X —are in $p\mathcal{E}^*$ and all named subsets of E are in $\mathcal{E}^e$ unless explicitly stated otherwise. We use the American convention for limits. For example, if f is a numerical function on $\mathbb{R}$, we write $\lim_{s \downarrow t} f(s)$ rather than $\lim_{s \downarrow\downarrow t} f(s)$. However, we use the notation $t_n \downarrow\downarrow t$ to denote a sequence (t_n) with $t_n > t$ for each n and which decreases to t. We write $\uparrow \lim_{s \downarrow t} f(s) = L$ to indicate that $f(s)$ increases to L as s decreases to t, etc. We use the standard notation $\mathbb{Q}$ and $\mathbb{R}$ for the rationals and the reals respectively, while $\mathbb{Q}^+$, $\mathbb{R}^+$, and $\mathbb{R}^{++}$ denote $\mathbb{Q} \cap [0, \infty[$, $\mathbb{R} \cap [0, \infty[$ and $\mathbb{R} \cap]0, \infty[$ respectively. $\mathcal{B}$, $\mathcal{B}^+$, and $\mathcal{B}^{++}$ denote the Borel σ-algebras of $\mathbb{R}$, $\mathbb{R}^+$, and $\mathbb{R}^{++}$ respectively. As usual in the theory of Markov processes almost surely (a.s.) means a.s. P^μ for each initial measure μ on E. Similarly indistinguishable means P^μ indistinguishable for each μ.

There is one technical convention that we adopt deserving special mention. If $f \in p\mathcal{E}^*$, then one readily checks that $(t, \omega) \to f \circ X_t(\omega)$ is $(\mathcal{B}^+ \otimes \mathcal{F}^0)^*$ measurable. In particular if P is *any* probability on Ω, $t \to f \circ X_t(\omega)$ is Lebesgue measurable for P a.e. ω. Consequently integrals of the form $\int \varphi(t) f \circ X_t\,dt$ for $\varphi \in p\mathcal{B}^+$ are defined P a.e., and since P is arbitrary they are $\mathcal{F}^*$ measurable provided they are defined arbitrarily (or left undefined) on the set of ω's such that $t \to f \circ X_t(\omega)$ is not Lebesgue measurable. If we want or need an unambiguously defined function of ω we replace the integral by the outer integral, $\int^* \varphi(t) f \circ X_t(\omega)\,dt$, discussed in Appendix B. The reader should consult Appendix B for more details. In the sequel we shall use the symbol $\int \varphi(t) f \circ X_t\,dt$ in this, perhaps, somewhat ambiguous manner.

2. Excessive Measures

In this section we shall define several important subclasses of excessive measures and detail two Riesz type decompositions of excessive measures.

A σ-finite measure ξ on E is *invariant* provided $\xi = \xi P_t$ for each $t > 0$. We write Inv for the class of invariant measures. Clearly Inv $\subset$ Exc. We write Inv (X) or Inv (P_t) if we want to emphasize the process X or the semigroup (P_t). This same notational scheme will be used for the other classes of excessive measures to be introduced in this section. If μ is a measure on E, then it is easily checked that $(\mu U)P_t \leq \mu U$ and so $\mu U \in$ Exc if and only if it is σ-finite. Excessive measures of the form μU are called *potentials* and we write Pot for the class of potentials. If $f > 0$ then $Uf > 0$, and so if $\mu U \in$ Pot then μ must be σ-finite. The converse is false even if U is proper; for example, consider Brownian motion in three or more dimensions and take μ to be Lebesgue measure. On the other hand if U is proper and μ is a finite measure, then $\mu U \in$ Pot.

Suppose $\mu U \in$ Pot and $f \geq 0$ with $\mu U(f) < \infty$. Then

$$\mu U P_t(f) = \int_t^\infty \mu P_s(f)\, ds \to 0 \quad \text{as} \quad t \to \infty .$$

This leads to the following definition. An excessive measure ξ is *purely excessive* provided $\xi P_t(f) \to 0$ as $t \to \infty$ whenever $f \geq 0$ with $\xi(f) < \infty$. The class of purely excessive measures is denoted by Pur. Then Pot $\subset$ Pur $\subset$ Exc.

We need some elementary facts about measures. The standard proofs are left to the reader. If μ and ν are measures on E we write $\mu \leq \nu$ provided $\mu(B) \leq \nu(B)$ for all $B \in \mathcal{E}$ and, hence all $B \in \mathcal{E}^*$. If μ and ν are σ-finite measures and $\mu \leq \nu$, then there exists a unique σ-finite measure, λ, such that $\mu + \lambda = \nu$. We write $\lambda = \nu - \mu$. If (μ_n) is a decreasing sequence of σ-finite measures, then there exists a unique σ-finite measure μ such that $\mu_n(B) \downarrow \mu(B)$ whenever $\mu_n(B) < \infty$ for some n. Then $\mu_n(f) \downarrow \mu(f)$ if $f \geq 0$ and $\mu_n(f) < \infty$ for some n. We write $\mu = \downarrow \lim \mu_n$ or simply $\mu_n \downarrow \mu$. This extends to an indexed family $(\mu_t)_{t>0}$ of σ-finite measures which is decreasing ($\mu_s \geq \mu_t$ if $s \leq t$). Note that if (μ_n) (or (μ_t)) is decreasing and $\mu_n(f) \to 0$ for a single $f > 0$ with $\mu_n(f) < \infty$ for some n, then $\mu_n \downarrow 0$.

We come now to the first decomposition of excessive measures.

(2.1) **Theorem.** *Let $\xi \in$ Exc. Then ξ may be written uniquely as $\xi = \xi_i + \xi_p$ where $\xi_i \in$ Inv and $\xi_p \in$ Pur. Moreover $\xi_i = \downarrow \lim_{t\to\infty} \xi P_t$.*

Proof. Since $\xi \in$ Exc, $(\xi P_t)_{t>0}$ is decreasing and so $\xi_i := \lim_{t\to\infty} \xi P_t$ exists as a σ-finite measure and $\xi_i \leq \xi$. If $f \geq 0$ with $\xi(f) < \infty$ and $s > 0$, then $\xi_i(P_s f) = \downarrow \lim_{t\to\infty} \xi P_t(P_s f) = \xi_i(f)$. Consequently $\xi_i \in$ Inv. Now defining $\xi_p := \xi - \xi_i$, it follows from the invariance of ξ_i that $\xi_p \in$ Exc and, hence, $\xi_p \in$ Pur. If $\xi = \eta + \lambda$ with $\eta \in$ Inv and $\lambda \in$ Pur, then since $\lambda P_t \downarrow 0$ as $t \to \infty$ it follows that $\xi_i = \eta$, and then that $\xi_p = \lambda$. ∎

The next decomposition is considerably less elementary than the one detailed in Theorem 2.1. It is due to Dynkin [**DY80**]. However, we shall follow Blumenthal [**B86**] in our discussion. We begin with an elementary fact.

(2.2) **Proposition.** *Let $\xi \in$ Pur. Then there exists an increasing sequence of potentials $(\mu_n U)$ with $\mu_n U \uparrow \xi$.*

Proof. Since $\xi \in$ Exc we may define σ-finite measures $\mu_n := n[\xi - P_{1/n}\xi]$. Let $f \geq 0$ with $\xi(f) < \infty$. Then

$$(2.3)\ \mu_n U(f) =\uparrow \lim_{a\to\infty} \int_0^a \mu_n P_t(f)\,dt$$

$$=\uparrow \lim_{a\to\infty} \left[n\int_0^{1/n} \xi P_t(f)\,dt - n\int_a^{a+1/n} \xi P_t(f)\,dt \right]$$

$$= n\int_0^{1/n} \xi P_t(f)\,dt\,,$$

because $\xi P_a(f) \downarrow 0$ as $a \to \infty$. Hence

$$\mu_n U = n\int_0^{1/n} \xi P_t\,dt = \int_0^1 \xi P_{t/n}\,dt \uparrow \xi$$

as $n \to \infty$. ■

We now introduce two additional classes of excessive measures. A measure $\xi \in \mathrm{Exc}$ is *conservative* provided $\mu U \leq \xi$ implies that $\mu U = 0$, while $\xi \in \mathrm{Exc}$ is *dissipative* provided $\xi = \sup\{\mu U\colon \mu U \leq \xi\}$. Naturally the class of conservative (resp. dissipative) measures is denoted by Con (resp. Dis). Of course, the supremum in the definition of a dissipative measure is the supremum relative to the order "$\leq$" in the class of σ-finite measures which is a boundedly complete lattice relative to this order. It follows from (2.2) that Pur $\subset$ Dis. We now may state the main result of this section.

(2.4) **Theorem.** *Let $\xi \in \mathrm{Exc}$. Then ξ may be written uniquely as $\xi = \xi_c + \xi_d$ where $\xi_c \in \mathrm{Con}$ and $\xi_d \in \mathrm{Dis}$. If $g > 0$ and $\xi(g) < \infty$, then ξ_c (resp. ξ_d) is the restriction of ξ to $\{Ug = \infty\}$ (resp. $\{Ug < \infty\}$). If $u \in S$, then for each t, $u = P_t u$ a.e. ξ_c.*

We shall break up the proof of Theorem 2.4 into a series of propositions.

(2.5) **Proposition.** Con $\subset$ Inv.

Proof. Let $\xi \in \text{Con}$ and let $\xi = \xi_i + \xi_p$ be the decomposition of ξ in Theorem 2.1. If $\mu U \leq \xi_p \leq \xi$, then $\mu U = 0$. Consequently, using (2.2), $\xi_p = 0$. ■

(2.6) **Definition.** A set $D \subset E$ is *dissipative* provided $D \in \mathcal{E}^e$, D is finely open, and there exists $h \in p\mathcal{E}^e$ with $h > 0$ on D and $Uh \leq 1$ everywhere.

Obviously a countable union of dissipative sets is dissipative and so is a finely open $\mathcal{E}^e$ measurable subset of a dissipative subset. The next proposition gives examples of dissipative sets.

(2.7) **Proposition.** (i) *If* $u \in S$ *and* $t > 0$, *then* $\{P_t u < u\}$ *is dissipative.* (ii) *If* $g \geq 0$, *then* $\{0 < Ug < \infty\}$ *is dissipative.*

Proof. Suppose first that $u \in S$ is bounded and that $P_r u \to 0$ as $r \to \infty$. Then as in the proof of (2.2), $U(u - P_t u) = \int_0^t P_r u\, dr$. Hence $h = (t\|u\|)^{-1}(u - P_t u) > 0$ on $D := \{P_t u < u\}$ and $Uh \leq 1$. Since $P_t u \in S$, $D \in \mathcal{E}^e$ and D is finely open. Hence D is dissipative. If $u \in S$ is bounded, then $v := \downarrow \lim_{r \to \infty} P_r u$ is in S and $P_r v = v$ for each $r \geq 0$. Therefore $w := u - v \in S$ and $P_r w \to 0$ as $r \to \infty$. But $D := \{P_t u < u\} = \{P_t w < w\}$ and so it is dissipative for any bounded $u \in S$. If $u \in S$ is arbitrary, then $u_n := u \wedge n \in S$ and $\{P_t u < u\} \subset \bigcup_n \{P_t u_n < u_n\}$, establishing (i). If $Ug(x) < \infty$, then $P_t Ug(x) \to 0$ as $t \to \infty$. Therefore $\{0 < Ug < \infty\} \subset \bigcup_n \{P_n Ug < Ug\}$, proving (ii). ■

We now fix $\xi \in \text{Exc}$. Let γ be a finite measure equivalent to ξ. Because dissipative sets are closed under countable unions, we may choose a dissipative set A of maximal γ measure. Then if B is dissipative, $\gamma(B \backslash A) = 0$ and hence $\xi(B \backslash A) = 0$. Let $h \in p\mathcal{E}^e$ with $h > 0$ on A and $Uh \leq 1$. Define

$$D = \{Uh > 0\}, \quad C = \{Uh = 0\}. \tag{2.8}$$

From (2.7ii), D is dissipative and since A is finely open, $A \subset D$ and so $\xi(D \backslash A) = 0$. Clearly if B is dissipative, then $\xi(B \cap C) = \xi(B \backslash D) = 0$. If λ is any measure on E and $B \in \mathcal{E}^*$, then λ_B denotes the restriction of λ to B. It will turn out that $\xi = \xi_C + \xi_D$ is the decomposition of Theorem 2.4. The next result is the key step in the proof.

(2.9) **Proposition.** $\xi_D \in$ Dis *and* $\xi_C \in$ Con. *If* $u \in S$ *then for each* $t > 0$, $P_t u = u$ *a.e.* ξ_C.

Proof. The last assertion is an immediate consequence of (2.7i) and the fact that if B is dissipative, then $\xi(B \cap C) = 0$. Because $C = \{Uh = 0\}$, C is an absorbing set, that is, if $x \in C$, $P^x\ (T_D < \infty) = 0$. In particular $P_t(x, \cdot)$ is carried by C if $x \in C$. Therefore it is easily checked that $\xi_C \in$ Exc. Suppose $\mu U \leq \xi_C$. If $\mu U \neq 0$, there exists B with $0 < \mu U(B) < \infty$. But then $\mu P_t U(B) \to 0$ as $t \to \infty$, and so for sufficiently large t, $\mu(\{P_t U 1_B < U 1_B\}) > 0$. Let H be the set in braces. It is dissipative and $\xi_C(H) \geq \mu U(H) > 0$ because, H being finely open, $U1_H > 0$ on H. But $\xi_C(H) = 0$ since H is dissipative. Therefore $\xi_C \in$ Con. Since $\xi_C \in$ Inv by (2.5) and $\xi = \xi_C + \xi_D$ it is immediate that $\xi_D \in$ Exc. It remains to show that $\xi_D \in$ Dis.

To this end choose an increasing sequence (A_n) of sets with $\xi(A_n) < \infty$ and $D = \cup A_n$. Fix $g \in p\mathcal{E}^e$ with $g > 0$ on D and $Ug \leq 1$. Let $\xi_n := \xi_{A_n}$ and $\mu_n := n\xi_n$. Each μ_n is finite and $\mu_n \leq n\xi_D$, so μ_n is carried by D. Since $\mu_n U^q \leq q^{-1} n \xi_D$, $\mu_n U^q(C) = 0$ and letting $q \to 0$ it follows that $\mu_n U(C) = 0$. Moreover for $\epsilon > 0$,

$$\epsilon \mu_n U(\{g > \epsilon\}) \leq \mu_n U g \leq \mu_n(1) < \infty .$$

Now $E = C \cup \bigcup_{n \geq 1} \{g > 1/n\}$ and so $\mu_n U$ is σ-finite, that is $\mu_n U \in$ Pot. Let η_n be the excessive measure $\mu_n U \wedge \xi_D$. Since $\mu_n U \in$ Pot, $\eta_n \in$ Pur. Let $\nu_n := n[\eta_n - P_{1/n}\eta_n]$. Define $\eta_{n,k} := k \int_0^{1/k} \eta_n P_t\, dt = \int_0^1 \eta_n P_{t/k}\, dt$. Since (η_n) is increasing,

$(\eta_{n,k})$ increases with both n and k, and by (2.3), $\nu_n U = \eta_{n,n}$. Consequently $(\nu_n U)$ increases, $\nu_n U \leq \eta_n \leq \xi_D$, and $\lim \nu_n U \geq \eta_k$ for each k. Thus to complete the argument, it suffices to show that $\eta_n \uparrow \xi_D$. Let θ_n be the measure $\int_0^1 \xi_n P_s\, ds$. Then (θ_n) is increasing and $\theta_n \leq \xi_D$. Let $f_n = d\theta_n/d\xi_D$. Then $f_1 \leq f_2 \leq \cdots \leq 1$ a.e. ξ_D. Now $\theta_n \uparrow \theta := \int_0^1 \xi_D P_s\, ds$ which is evidently equivalent to ξ_D. Therefore $\lim f_n > 0$ a.e. ξ_D. But the measure $n\theta_n \wedge \xi_D$ has derivative $(nf_n) \wedge 1$ with respect to ξ_D and this sequence of derivatives has limit 1 a.e. ξ_D. As a result $n\theta_n \wedge \xi_D \uparrow \xi_D$. But $\theta_n \leq \xi_n U$ and so $\eta_n = n\xi_n U \wedge \xi_D \uparrow \xi_D$. ∎

(2.10) **Remark.** Note that the proof of (2.9) shows that ξ_D is the increasing limit of the *sequence* of potentials $(\nu_n U)$ and $\nu_n \leq n\eta_n \leq n\xi_D$.

(2.11) **Proposition.** *Let $g > 0$ and $\xi(g) < \infty$. Then ξ_C and ξ_D are the restrictions of ξ to $\{Ug = \infty\}$ and $\{Ug < \infty\}$ respectively.*

Proof. Since $Ug > 0$, the set $\{Ug < \infty\}$ is dissipative by (2.7). Therefore $\xi_C(\{Ug < \infty\}) = 0$. To complete the proof of (2.11), it suffices to show that $\xi_D(\{Ug = \infty\}) = 0$, and, in light of (2.10), it suffices to show $\mu U(\{Ug = \infty\}) = 0$ whenever $\mu U \leq \xi_D$. But then $\mu U(g) \leq \xi(g) < \infty$, and so $\mu(\{Ug = \infty\}) = 0$. Since $\{Ug < \infty\}$ is absorbing and carries μ, it also carries μU. ∎

Combining (2.9) and (2.11) establishes all of the assertions in Theorem 2.4 except for the uniqueness of the decomposition. For the uniqueness suppose that $\xi = \eta + \theta$ with $\eta \in$ Con and $\theta \in$ Dis. Since C is absorbing, $\theta_C \in$ Exc. If $\mu U \leq \theta_C$, then $\mu U \leq \xi_C$ and hence $\mu U = 0$. Therefore $\theta_C \in$ Con. We claim that, in fact, $\theta_C = 0$. To this end first note that if $\lambda \in$ Pur and $\lambda \leq \theta_C$, then because of (2.2), $\lambda = 0$. Because $\theta = \sup\{\mu U : \mu U \leq \theta\}$ there exists a countable family

$(\mu_j U)$ of potentials such that $\lambda_n := \bigvee_{j \leq n} \mu_j U \uparrow \theta$. But then $\lambda_n \wedge \theta_C \uparrow \theta_C$ and since $\lambda_n \wedge \theta_C \in$ Pur, being dominated by $\left(\sum_{j \leq n} \mu_j\right) U$, $\theta_C = 0$. Therefore $\xi_C = \eta_C + \theta_C = \eta_C$. Hence $\eta_C \in$ Con $\subset$ Inv and so $\eta_D \in$ Exc and $\eta_D \leq \xi_D$. By (2.10) there exist $\nu_n U \uparrow \xi_D$ which implies $\nu_n U \wedge \eta_D \in$ Pur; consequently $\eta_D = 0$. Therefore $\xi_D = \theta_D$. As a result $\theta + \eta = \xi = \theta_D + \eta_C$ and since $\theta_D \leq \theta$, $\eta_C \leq \eta$, it follows that $\xi_D = \theta_D = \theta$ and $\xi_C = \eta_C = \eta$. This completes the proof of Theorem 2.4.

Remark. An immediate consequence of Theorem 2.4 is the fact that given ξ, $\eta \in$ Exc with $\eta \leq \xi$, then $\xi \in$ Dis implies $\eta \in$ Dis and $\xi \in$ Con implies $\eta \in$ Con. That is, the cones Dis and Con are *solid* relative to the order "$\leq$".

Next we shall develop several properties of the various classes of excessive measures we have defined. We begin with an important uniqueness result.

(2.12) **Theorem.** *If μU and νU are σ-finite and $\mu U = \nu U$, then $\mu = \nu$.*

Proof. Let $q > 0$. Then the resolvent equation implies $\mu U = \mu U^q + q\mu U U^q$. Since both measures on the right side of this equality are σ-finite, μU^q is uniquely determined by μU. Therefore $\mu U^q = \nu U^q$ for all $q > 0$. Fix $q > 0$. Because U^q is proper and μU^q is σ-finite we may choose $g > 0$ with $h := U^q g \leq 1$ and $\mu(h) = \mu U^q(g) < \infty$. Let $u \in bS^q$. Then for $r > 0$,

$$r\mu U^{q+r}(uh) = r\nu U^{q+r}(uh).$$

It is elementary that the product of two elements in bS^q is in $bS^q - bS^q$. See, for example, [**S**, (8.7)]. Writing $uh = u_1 - u_2$ with $u_1, u_2 \in bS^q$ we see that $rU^{q+r}(uh)$ approaches uh as $r \to \infty$ and is bounded by $\|u\|h$. Consequently letting $r \to \infty$ in the above display yields $\mu(uh) = \nu(uh)$ for $u \in bS^q$, and hence for $u \in bS^q - bS^q$. Since $bS^q - bS^q$ is an algebra and

$\mathcal{E} \subset \mathcal{E}^e = \sigma(bS^q)$, the monotone class theorem [**DM**, I-21] implies that the finite measures $h\mu$ and $h\nu$ are equal. Since $0 < h \leq 1$, this gives $\mu = \nu$. ∎

(2.13) **Remark.** Note that the proof of (2.12) is valid for any strictly positive subMarkovian resolvent (U^q) on $(E, \mathcal{E})$ as defined in [**DM**, XII-7] for which $\mathcal{E} \subset \sigma(S^q)$ for some $q > 0$.

If X is transient (i.e., U is proper), then every excessive measure is dissipative. In the general case if $\xi \in \text{Dis}$ (resp. $\xi \in \text{Con}$), then X "relative to ξ" behaves like a transient (resp. recurrent) process. For example if $\xi \in \text{Dis}$ and $g > 0$ with $\xi(g) < \infty$, then (2.4) states that ξ is carried by the absorbing set $A = \{Ug < \infty\}$. One may restrict X to A and the resolvent and semigroup of this restricted process X^A are just the restrictions of (U^q) and (P_t) to A, and X^A is a transient right process with state space A. Thus if $u \in S$, the restriction of u to A, u_A, is in $S(X^A)$. Consequently there exists a sequence (f_n) of positive functions on A such that $Uf_n \uparrow u$ on A. Extending each f_n to E by setting it equal to zero on $E \backslash A$ and using the fact that for each $x \in E \backslash A$, T_A may be approximated by the hitting times of compact subsets a.s. P^x, it follows that (Uf_n) increases on all of E and has limit u on $A = \{Ug < \infty\}$. Note also that if $\mu U \leq \xi$, then $\mu U(g) < \infty$ and so $\mu(\{Ug = \infty\}) = 0$. Thus we have established the following result.

(2.14) **Proposition.** *If $\xi \in \text{Dis}$ and $u \in S$, then there exists an increasing sequence (Uf_n) of potentials such that $\lim Uf_n = u$ a.e. ξ and also a.e. μ whenever $\mu U \leq \xi$.*

The next result is an immediate consequence of (2.14).

$$\text{(2.15)} \qquad \nu U \leq \mu U \in \text{Pot} \Rightarrow \nu(u) \leq \mu(u) \quad \text{for all} \quad u \in S.$$

The final result of this section is a strengthening of the last assertion in Theorem 2.4. It should be compared to the well-known fact that an excessive function of a recurrent process is constant. See [**G80**].

(2.16) **Theorem.** *Let* $\xi \in \mathrm{Con}$ *and* $u \in S$. *Then*

$$P^{\xi}[u \circ X_t \neq u \circ X_0 \quad \text{for some} \quad t \geq 0] = 0\,.$$

Proof. If $B \in \mathcal{E}^e$ let $L_B = \sup\{t \geq 0 : X_t \in B\}$ where the supremum of the empty set is zero. Then $L_B \in \mathcal{F}$ and $\psi_B(x) := P^x[0 < L_B < \infty]$ is excessive. Now $P_t\psi_B(x) = P^x[t < L_B < \infty] \to 0$ as $t \to \infty$ and consequently because $\xi \in \mathrm{Con}$, Theorem 2.4 implies that $\psi_B = 0$ a.e. ξ. Also $P^x(L_B > 0) = P^x(T_B < \infty)$ for any $B \in \mathcal{E}^e$. In proving (2.16) it suffices to assume u bounded. Now define $A_t = \{u > t\}$, $B_t = \{u < t\}$, and $F = \cap\{\psi_{A_r} = 0 = \psi_{B_q}\}$ where the intersection is over all positive rationals r and q. Then $\xi(F^c) = 0$. Suppose $x \in F$ and $q < r < u(x)$. Then $x \in A_r$ and since A_r is finely open $P^x(L_{A_r} > 0) = 1$. Hence $P^x(L_{A_r} = \infty) = 1$; that is, a.s. P^x, $u \circ X_t > r$ for arbitrarily large t. If $P^x(T_{B_q} < \infty) > 0$, then $P^x(L_{B_q} = \infty) > 0$ and so $u \circ X_t < q$ for arbitrarily large t with positive P^x probability. Since u is bounded, $\lim_{t\to\infty} u \circ X_t$ exists a.s. P^x and therefore $P^x(T_{B_q} < \infty) = 0$. Since $q < u(x)$ is arbitrary one has a.s. P^x, $u \circ X_t \geq u(x)$ for all t. Similarly a.s. P^x, $u \circ X_t \leq u(x)$ for all t. But $x \in F$ is arbitrary and $\xi(F^c) = 0$ which gives (2.16). ∎

(2.17) **Remark.** Let us make explicit the result contained in (2.10). If $\xi \in \mathrm{Dis}$, then there exists a sequence of potentials $(\mu_n U)$ increasing to ξ with $\mu_n \leq n\xi$ for each n. Moreover making use of the device of restricting to the absorbing set $\{Ug < \infty\}$ where $g > 0$ with $\xi(g) < \infty$ used in proving (2.14), it follows from [**DM**, XII-38] that one may also suppose that each μ_n is finite. Similarly it follows from [**DM**, XII-8] that there exist an $f > 0$ with $\xi(f) < \infty$ and $Uf \leq 1$ a.e. ξ. See also [**G80**] in this connection.

(2.18) **Remark.** By working with the q-subprocess of X all of these results apply to q-excessive measures. Of course, since U^q is proper if $q > 0$, $\mathrm{Exc}^q = \mathrm{Dis}^q$ using a self-explanatory notation. However there may exist non-trivial q-invariant measures. For example, if X is translation to the right at speed one on $\mathbb{R}$, then $\xi(dx) = e^{-qx}\,dx$ is q-invariant.

3. The Energy Functional

In [**Me73**] (see also [**DM**, XII-39]) Meyer associated with each $\xi \in \text{Exc}$ and $u \in S$ a number $L(\xi, u)$ with $0 \leq L(\xi, u) \leq \infty$, which generalizes the notion of energy in classical potential theory. We shall proceed in a slightly different manner from that used in [**DM**, XII-39].

(3.1) **Definition.** The *energy functional* L on $\text{Exc} \times S$ is defined by

$$L(\xi, u) = \sup \{\mu(u) \colon \mu U \leq \xi\}.$$

The following facts are immediate consequences of the definition.

(3.2) (a) $\xi \in \text{Con} \Longrightarrow L(\xi, \cdot) = 0$.

(b) $\xi_1 \leq \xi_2 \Longrightarrow L(\xi_1, \cdot) \leq L(\xi_2, \cdot)$.

(c) $\mu U \in \text{Pot} \Longrightarrow L(\mu U, u) = \mu(u)$.

Of course in (3.2b), ξ_1 and ξ_2 are excessive and in (3.2c), $u \in S$. We won't repeat such qualifying remarks again since L is defined on $\text{Exc} \times S$ only. If $g > 0$ with $\xi(g) < \infty$ and $\mu U \leq \xi$, then μU being dissipative implies that μU is carried by $\{Ug < \infty\}$ and so $\mu U \leq \xi_d$. See Theorem 2.4. As a result

$$L(\xi, \cdot) = L(\xi_d, \cdot). \tag{3.3}$$

We collect some of the properties of the functional L in the following proposition.

(3.4) **Proposition.** *In the following ξ and u with or without subscripts denote elements of* Exc *and* S *respectively.*

(i) $u_1 \leq u_2$ *a.e.* $\xi \Longrightarrow L(\xi, u_1) \leq L(\xi, u_2)$.

(ii) $u_n \uparrow u$ *a.e.* $\xi \Longrightarrow L(\xi, u_n) \uparrow L(\xi, u)$.

(iii) $L(\xi, Uf) = \xi_d(f)$.

(iv) $\xi_n \uparrow \xi \Longrightarrow L(\xi_n, \cdot) \uparrow L(\xi, \cdot)$.

(v) *If* $\alpha_1, \alpha_2 \geq 0$, *then* $L(\xi, \alpha_1 u_1 + \alpha_2 u_2) = \alpha_1 L(\xi, u_1) + \alpha_2 L(\xi, u_2)$ *and* $L(\alpha_1 \xi_1 + \alpha_2 \xi_2, u) = \alpha_1 L(\xi_1, u) + \alpha_2 L(\xi_2, u)$.

(vi) $L(\xi, u) = 0 \Longleftrightarrow u = 0$ *a.e* ξ_d.

(vii) $L(\xi_1, \cdot) = L(\xi_2, \cdot) \Longleftrightarrow (\xi_1)_d = (\xi_2)_d$.

Proof. (i) Suppose $\mu U \leq \xi$. Then $\mu U(\{u_1 > u_2\}) = 0$. But $\{u_1 > u_2\}$ is finely open and so $\mu(\{u_1 > u_2\}) = 0$. Hence $\mu(u_1) \leq \mu(u_2)$, proving (i).

Before proceeding with the proof we require a special case of (iv).

(3.5) **Lemma.** *If* $\mu_n U \uparrow \xi$, *then* $L(\mu_n U, \cdot) \uparrow L(\xi, \cdot)$.

Proof. By (3.2b), $L(\mu_n U, \cdot)$ increases to a limit that is no greater than $L(\xi, \cdot)$. The hypothesis implies that $\xi \in \text{Dis}$, and so, given $u \in S$, by (2.14) there exists an increasing sequence (Uf_k) such that $Uf_k \uparrow u$ a.e. ξ and also a.e. ν whenever $\nu U \leq \xi$. Fix $\mu U \leq \xi$. Then

$$\begin{aligned}\mu(u) &= \uparrow \lim_k \mu(Uf_k) \leq \uparrow \lim_k \xi(f_k) \\ &= \uparrow \lim_k \uparrow \lim_n \mu_n U(f_k) = \uparrow \lim_n \mu_n(u) = \uparrow \lim_n L(\mu_n U, u),\end{aligned}$$

where the next to last equality follows because $\mu_n U \leq \xi$ and last follows from (3.2c). Since $\mu U \leq \xi$ was arbitrary this establishes (3.5). ∎

We now return to the proof of (3.4).

(ii) In view of (3.3) it suffices to prove this when $\xi \in \text{Dis}$. By (2.17) there exist potentials $\mu_n U \uparrow \xi$. Note that (i) and (3.2c) imply that $\mu_n(u_k) = L(\mu_n U, u_k) \leq L(\mu_n U, u_{k+1}) = \mu_n(u_{k+1})$. Let $v = \liminf u_k$. Then $v = u$ a.e. ξ, v is supermedian and if $\bar{v}$ is the excessive regularization of v, then $\bar{v} \leq v$ and $\{\bar{v} < v\}$ is of potential zero. But $q\xi U^q \uparrow \xi$ as

$q \to \infty$ and so $\xi(\{\bar{v} < v\}) = 0$. Therefore $\bar{v} = u$ a.e. ξ and a.e. μ_n for each n since $\{\bar{v} \neq u\}$ is finely open. Now

$$\begin{aligned} L(\mu_n U, u) &= \mu_n(u) = \mu_n(\bar{v}) \leq \mu_n(v) \leq \liminf_k \mu_n(u_k) \\ &= \uparrow \lim_k L(\mu_n U, u_k), \end{aligned}$$

where we have used Fatou's lemma and the fact that $k \to \mu_n(u_k)$ is increasing. Using (3.5) and the fact that $L(\mu_n U, u_k)$ increases with both n and k we see that $L(\xi, u) \leq \lim_k L(\xi, u_k)$. Because the reverse inequality follows from (i), this yields (ii).

(iii) Again we may suppose $\xi \in \text{Dis}$. Let $\mu_n U \uparrow \xi$. Using (3.5) we find

$$L(\xi, Uf) = \lim L(\mu_n U, Uf) = \lim \mu_n(Uf) = \xi(f).$$

(iv) Let $g > 0$ with $\xi(g) < \infty$. Since ξ_d and $(\xi_n)_d$ are the restrictions of ξ and ξ_n respectively to $\{Ug < \infty\}$ it follows that $(\xi_n)_d \uparrow \xi_d$. Hence we may suppose $\xi \in \text{Dis}$ in proving (iv). Let $u \in S$ and choose an increasing sequence (Uf_k) of potentials with $\lim Uf_k = u$ a.e. ξ and a.e. μ whenever $\mu U \leq \xi$. If $\mu U \leq \xi$, then

$$\begin{aligned} \mu(u) &= \lim_k \mu(Uf_k) \leq \lim_k \lim_n \xi_n(f_k) \\ &= \lim_k \lim_n L(\xi_n, Uf_k) = \lim_n L(\xi_n, u) \end{aligned}$$

where we have used (iii) for the next to last equality, and (ii) and the fact that $L(\xi_n, Uf_k)$ increases with both n and k for the last. It is now immediate that $L(\xi, u) \leq \lim L(\xi_n, u)$ and so (iv) follows from (3.2b).

(v) For the first assertion one may suppose $\xi \in \text{Dis}$, in which case it is an immediate consequence of (ii), (iii), and (2.14). Since $(\alpha_1\xi_1 + \alpha_2\xi_2)_c = \alpha_1(\xi_1)_c + \alpha_2(\xi_2)_c$ we may suppose $\xi_1, \xi_2 \in \text{Dis}$ in checking the second assertion. But then it follows from (3.5) and (3.2c).

(vi) Again we may assume $\xi \in \mathrm{Dis}$. Suppose $L(\xi, u) = 0$. Let $Uf_n \uparrow u$ a.e. ξ. Then from (ii) and (iii), $\xi(f_n) = 0$ for each n. If $q > 0$, $q\xi U^q f_n \le \xi(f_n)$ and so $U^q f_n = 0$ a.e. ξ. But $U^q f_n \uparrow Uf_n$ as $q \downarrow 0$. Therefore $Uf_n = 0$ a.e. ξ, and hence, $u = 0$ a.e. ξ. The reverse implication is clear.

(vii) This is clear in light of (iii) and (3.3). ∎

Suppose X and $\hat{X}$ are in strong duality with respect to a σ-finite measure m as in Chapter VI of [**BG**]. Let $u(x,y)$ be the potential density kernel. If $h \in S$ has the form $h(x) = \int u(x,y)\nu(dy) := U\nu(x)$ and $\mu U \in \mathrm{Pot}$, then

$$L(\mu U, h) = \mu(h) = \iint \mu(dx)u(x,y)\nu(dy)$$

which represents the "mutual energy" of the measures μ and ν in potential theory. This is the reason I call L the energy functional. In general one should think of $L(\xi, u)$ as the "energy" between an excessive measure ξ and an excessive function u. In [**DM**], L is called the mass functional. Because of Einstein, this is just a choice of units.

The next result shows that our definition agrees with that in [**DM**, XII-39]. Recall $\langle \lambda, f\rangle = \int f\, d\lambda$.

(3.6) **Proposition.** (i) *If $\xi \in \mathrm{Pur}$ and $u \in S$, then*

$$L(\xi, u) = \uparrow \lim_{t \to 0} t^{-1}\langle \xi - \xi P_t, u\rangle = \uparrow \lim_{q \to \infty} q\langle \xi - q\xi U^q, u\rangle .$$

(ii) *If $\xi \in \mathrm{Exc}$ and $u \in S$ is such that a.e. ξ, $u < \infty$ and $P_t u \to 0$ as $t \to \infty$, then*

$$L(\xi, u) = \uparrow \lim_{t \to 0} t^{-1}\langle \xi, u - P_t u\rangle = \uparrow \lim_{q \to \infty} q\langle \xi, u - qU^q u\rangle .$$

Proof. Suppose $\xi \in \mathrm{Pur}$. The proof of (2.2) shows that defining $\mu_t := t^{-1}[\xi - P_t\xi]$ one has $\mu_t U \uparrow \xi$ as $t \downarrow 0$. The first equality in (3.6i) now follows from (3.2c) and (3.4iv). As σ-finite measures

$$\xi - q\xi U^q = q \int_0^\infty e^{-qt}(\xi - \xi P_t)\, dt\,,$$

and so

$$q\langle \xi - q\xi U^q, u\rangle = \int_0^\infty e^{-t}(t/q)^{-1}\langle \xi - \xi P_{t/q}, u\rangle t\, dt\,.$$

Letting $q \to \infty$ the second equality in (3.6i) follows from the first. For (3.6ii), note first that $u - P_t u$ and $u - qU^q u$ are well defined a.e. ξ since $\xi(\{u = \infty\}) = 0$. It is well-known and easily checked that $t^{-1}U(u - P_t u)(x)$ exists and increases to $u(x)$ for each x such that $u(x) < \infty$ and $P_t u(x) \to 0$ as $t \to \infty$. Define $u_t(x) = t^{-1}(u - P_t u)(x)$ if $u(x) < \infty$ and $u_t(x) = 0$ if $u(x) = \infty$. Since $\{u < \infty\}$ is absorbing $Uu_t \uparrow u$ a.e. ξ as $t \downarrow 0$. The first equality in (3.6ii) now follows from (3.4ii), (3.4iii), and the last assertion in Theorem 2.4. The second equality in (3.6ii) follows from the first as before. ∎

(3.7) **Remark.** One must be careful not to write the integrals appearing in (3.6) as differences. For example, it is *not* the case that

$$L(\xi, u) = \lim_{t \to 0} t^{-1}[\xi(u) - \xi P_t(u)]$$

even if ξ is a potential. Consider Brownian motion in $\mathbb{R}^3$ and let $\xi = \epsilon_0 U$. Then $\xi(1) = \xi P_t(1) = \infty$ while $L(\xi, 1) = 1$. Also these formulas are not valid, in general, when the hypotheses are not satisfied. For example, if X is Brownian motion in $\mathbb{R}^3$ and ξ is Lebesgue measure, then $\xi P_t = \xi$ for all t, but $L(\xi, \cdot) \neq 0$.

Of course all of the preceding results may be applied to the q-subprocess, X^q, of X. We let L^q denote the energy functional of X^q. Note that $\mathrm{Exc}(X^q) = \mathrm{Exc}^q(X)$ and $S(X^q) = S^q(X)$. If $0 \leq r < q$ then $\mathrm{Exc}(X^r) \subset \mathrm{Exc}(X^q)$ and $S(X^r) \subset S(X^q)$.

(3.8) **Proposition.** *Let* $0 \le r < q$ *and* $\xi \in \mathrm{Exc}(X^r)$, $u \in S(X^r)$. *Then*

$$L^q(\xi, u) = L^r(\xi, u) + (q - r)\xi(u).$$

Proof. Replacing X by X^r it suffices to prove this when $r = 0$. Moreover we may consider the cases $\xi \in \mathrm{Con}$ and $\xi \in \mathrm{Dis}$ separately. Suppose $\xi \in \mathrm{Con}$. Then $L(\xi, u) = 0$ and since $\mathrm{Con} \subset \mathrm{Inv}$, $\xi = q\xi U^q$. That is, ξ is the q-potential of $q\xi$ and so $L^q(\xi, u) = q\xi(u)$ by (3.2c) applied to X^q. Thus (3.8) holds in this case. Next suppose $\xi \in \mathrm{Dis}$. Then there exist potentials $\mu_n U$ increasing to ξ. Define $\nu_n := \mu_n + q\mu_n U$. Then by the resolvent equation $\nu_n U^q = \mu_n U \uparrow \xi$. Hence

$$L^q(\xi, u) = \uparrow \lim \nu_n(u) = \uparrow \lim \left[\mu_n(u) + q\mu_n U(u)\right].$$

But $\mu_n(u) \uparrow L(\xi, u)$ and $\mu_n U \uparrow \xi$ establishing (3.8) for $\xi \in \mathrm{Dis}$. ∎

(3.9) **Corollary.** *Let* $\xi \in \mathrm{Inv}$ *and* $u \in S$. (i) *If* $\xi(u) < \infty$, *then* $L(\xi, u) = 0$. (ii) *If* $\xi(u) < \infty$ *and* $u > 0$ *a.e.* ξ, *then* $\xi \in \mathrm{Con}$. *In particular, a finite invariant measure is conservative.*

Proof. Fix $q > 0$. Since $\xi \in \mathrm{Inv}$, $\xi = q\xi U^q$ and $L^q(\xi, u) = q\xi(u)$. Now (i) follows from (3.8). But then (3.4vi) states that $u = 0$ a.e. ξ_d which implies (ii). ∎

4. Balayage of Excessive Measures

Recall [**S**, (12.1)] that an $(\mathcal{F}_t)$ stopping time, T, is a *weak terminal time* provided (i) for each t, $t + T \circ \theta_t = T$ a.s. on $\{t < T\}$, and that T is *exact* if, in addition, (ii) $t_n + T \circ \theta_{t_n} \downarrow T$ a.s. whenever $t_n \downarrow\downarrow 0$. If T is an exact weak terminal time, then according to [**S**, (55.20)] there exists an $(\mathcal{F}_t)$ stopping time S with $S = T$ a.s. and such that (iii) $t + S(\theta_t \omega) = S(\omega)$ for all t, ω with $t < S(\omega)$ and (iv) $S(\omega) = \downarrow \lim_{t\downarrow 0} [t + S(\theta_t \omega)]$ for every ω. It follows from (iii) and (iv) that $t \to t + S(\theta_t \omega)$ is right continuous and increasing on $[0, \infty[$ for every ω. For simplicity we define a *terminal time* S to be an $(\mathcal{F}_t)$ stopping time that satisfies (iii), and we say that S is *exact* if, in addition, it satisfies (iv). In this language, [**S**, (55.20)] states that an exact weak terminal time is equal a.s. to an exact terminal. Note that if S is a terminal time, then (iii) implies that $t \to t + S \circ \theta_t$ is increasing and one readily checks that $S^* := \downarrow \lim_{t\downarrow 0} (t + S \circ \theta_t)$ is an exact terminal time called the *exact regularization* of S. Of course, if $B \in \mathcal{E}^e$ then $T_B := \inf\{t > 0 : X_t \in B\}$ is an exact terminal time in the above sense, while $D_B := \inf\{t \geq 0 : X_t \in B\}$ is a terminal time. Moreover T_B is the exact regularization of D_B.

If T is any $(\mathcal{F}_t)$ stopping time and $q \geq 0$, define for $f \in p\mathcal{E}^*$

$$P_T^q f(x) = P^x[e^{-qT} f \circ X_T]. \tag{4.1}$$

Obviously if $S = T$ a.s., then $P_T^q(x, \cdot) = P_S^q(x, \cdot)$ for each $x \in E$ and $q \geq 0$. Thus when dealing with an exact weak terminal time, T, these operators are unchanged if we replace T by an exact terminal time $S = T$ a.s. In what follows it is only these operators P_T^q that will play a role and so there is no loss of generality in supposing that T is an exact terminal time. We write P_T when $q = 0$. We also write P_B^q in place of $P_{T_B}^q$ as usual. Of course, P_T^q defines a subMarkovian kernel on $(E, \mathcal{E}^*)$. If $f \in S^q$ and T is a terminal time, then $u := P_T^q f$ is q-supermedian while if T is an exact terminal time P_T^q maps S^q into itself. We are going to define for T an exact terminal time an operator R_T^q on Exc^q that is dual to P_T^q relative to the pairing $L^q(\cdot,\cdot)$ between Exc^q and S^q. Actually R_T^q may be defined when T is replaced by an exact multiplicative functional, but we shall only consider exact terminal times. We refer the interested reader to **[FG89]** for the general case.

We now fix an exact terminal time T for the remainder of this section. If $q > 0$ and $\xi \in \mathrm{Exc}^q$, we define

$$(4.2) \qquad R_T^q \xi(f) := L^q(\xi, P_T^q U^q f).$$

In view of (3.4v) and (3.4ii)—applied, of course, to X^q— it is clear that $R_T^q \xi$ is a measure. Since $P_T^q U^q \le U^q$ and $\mathrm{Exc}^q = \mathrm{Dis}^q$ when $q > 0$, it follows that $R_T^q \xi \le \xi$. In particular $R_T^q \xi$ is σ-finite. Moreover, $U^q P_t^q = P_t^q U^q \le U^q$ and so $(R_T^q \xi) P_t^q \le R_T^q \xi$. That is $R_T^q \xi \in \mathrm{Exc}^q$. (Recall $P_t^q = e^{-qt} P_t$.) Here is another description of R_T^q which will prove useful. Let $\mu_n U^q \uparrow \xi$. Since $\xi \in \mathrm{Exc}^q = \mathrm{Dis}^q$ such a sequence of q-potentials always exists. Using (3.2c) and (3.4iv) we have

$$(4.3) \qquad \mu_n U^q \uparrow \xi \Longrightarrow \mu_n P_T^q U^q \uparrow R_T^q \xi.$$

(4.4) **Lemma.** *If $\xi \in \mathrm{Exc}$, then $q \to R_T^q \xi$ is decreasing on $]0, \infty[$.*

Proof. Fix $0 < r < q$. If $f \in b p\mathcal{E}$, a straightforward computation shows that

$$(4.5) \qquad U^r P_T^q U^q f(x) = P^x \int_0^\infty e^{(q-r)t} \int_{t+T\circ\theta_t}^\infty e^{-qs} f \circ X_s \, ds \, dt.$$

But $t + T \circ \theta_t \geq T \vee t$ and so

$$\int_{t+T\circ\theta_t}^{\infty} e^{-qs} f \circ X_s \, ds \leq \int_t^{\infty} 1_{[T,\infty[}(s) e^{-qs} f \circ X_s \, ds \, .$$

Substituting this into (4.5) and changing the order of integration we obtain

$$(4.6) \qquad (q-r) U^r P_T^q U^q \leq P_T^r U^r - P_T^q U^q \, .$$

Since $\xi \in \mathrm{Exc} \subset \mathrm{Dis}^r$ there exists a sequence $(\mu_n U^r)$ with $\mu_n U^r \uparrow \xi$. Setting $\nu_n := \mu_n + (q-r)\mu_n U^r$ we have $\nu_n U^q = \mu_n U^r \uparrow \xi$. Therefore using (4.3) and (4.6)

$$R_T^q \xi = \uparrow \lim \mu_n [I + (q-r) U^r] P_T^q U^q \leq \uparrow \lim \mu_n P_T^r U^r = R_T^r \xi \, ,$$

establishing (4.4). ∎

We now define for $\xi \in \mathrm{Exc}$

$$(4.7) \qquad R_T \xi = R_T^0 \xi = \uparrow \lim_{q \downarrow 0} R_T^q \xi \, .$$

Clearly $R_T \xi \leq \xi$ and $R_T : \mathrm{Exc} \to \mathrm{Exc}$. It is also immediate that $R_T(\alpha_1 \xi_1 + \alpha_2 \xi_2) = \alpha_1 R_T \xi_1 + \alpha_2 R_T \xi_2$ for $\xi_1, \xi_2 \in \mathrm{Exc}$ and $\alpha_1, \alpha_2 \geq 0$, and that $R_T \xi_1 \leq R_T \xi_2$ when $\xi_1 \leq \xi_2$. Moreover it follows from (4.2), (4.7), and (3.4i) that if T and S are exact terminal times and $T = S$ a.s., then $R_T \xi = R_S \xi$ for all $\xi \in \mathrm{Exc}$.

(4.8) **Proposition.** (i) *Let* $(\xi_n) \subset \mathrm{Exc}^q$ *with* $\xi_n \uparrow \xi \in \mathrm{Exc}^q$. *Then* $R_T^q \xi_n \uparrow R_T^q \xi$. (ii) *If* $\xi \in \mathrm{Dis}$ *and* $\mu_n U \uparrow \xi$, *then* $\mu_n P_T U \uparrow R_T \xi$.

Proof. For $q > 0$, (i) is immediate from (4.2) and (3.4iv). Suppose $(\xi_n) \in \mathrm{Exc}$ and $\xi_n \uparrow \xi \in \mathrm{Exc}$. By (4.4) and the case $q > 0$, $R_T^q \xi_n$ increases as n increases and as q decreases. Therefore

$$\uparrow \lim_n R_T \xi_n = \uparrow \lim_n \uparrow \lim_{q \downarrow 0} R_T^q \xi_n = \uparrow \lim_{q \downarrow 0} R_T^q \xi = R_T \xi \, .$$

This proves (i). Because of (i), to prove (ii) it suffices to show $R_T(\mu U) = \mu P_T U$ when $\mu U \in \text{Pot}$. But $\mu U = \mu(I + qU)U^q$ and so

$$(4.9) \qquad R_T(\mu U) = \lim_{q\to 0} R_T^q(\mu U) = \lim_{q\to 0} \mu(I + qU)P_T^q U^q .$$

Suppose $f \geq 0$ with $\mu U(f) < \infty$. Then $q\mu UU^q(f) = \mu U(f) - \mu U^q(f) \to 0$ as $q \to 0$. Since $q\mu UP_T^q U^q \leq q\mu UU^q$, it follows from (4.9) that $R_T(\mu U) = \mu P_T U$. ■

Perhaps it is worthwhile to record explicitly the following consequence of (4.8ii) and (4.3):

$$(4.10) \qquad \mu U^q \in \text{Pot}^q \Longrightarrow R_T^q(\mu U^q) = \mu P_T^q U^q; \quad q \geq 0 .$$

If $\xi \in \text{Exc}$, then $R_T\xi_c \leq \xi_c$ and $R_T\xi_d \leq \xi_d$. Therefore $R_T\xi_c \in \text{Con}$ and $R_T\xi_d \in \text{Dis}$ —see the remark above Theorem 2.12. By the uniqueness of the decomposition in Theorem 2.4 we obtain

$$(4.11) \quad \xi \in \text{Exc} \Longrightarrow (R_T\xi)_c = R_T\xi_c \quad \text{and} \quad (R_T\xi)_d = R_T\xi_d .$$

The next result establishes the "duality" between R_T^q and P_T^q.

(4.12) **Proposition.** *Let $q \geq 0$, $\xi \in \text{Exc}^q$, and $u \in S^q$. Then $L^q(R_T^q\xi, u) = L^q(\xi, P_T^q u)$. In particular if $\xi \in \text{Dis}$, $R_T\xi(f) = L(\xi, P_T Uf)$.*

Proof. Suppose $q = 0$. Because of (4.11) and (3.2a), it suffices to prove (4.12) for $\xi \in \text{Dis}$. Then there exist $\mu_n U \uparrow \xi$. By (4.8ii), $\mu_n P_T U \uparrow R_T\xi$. Therefore

$$\begin{aligned} L(R_T\xi, u) &= \lim_n L(\mu_n P_T U, u) = \lim_n \mu_n P_T u \\ &= \lim_n L(\mu_n U, P_T u) = L(\xi, P_T u) . \end{aligned}$$

Since $\text{Exc}^q = \text{Dis}^q$ when $q > 0$, this establishes the first claim in (4.12) for $q > 0$ also. The second assertion is an immediate consequence of the first and (3.4iii). ■

If $B \in \mathcal{E}^e$ and $\xi \in \text{Exc}^q$ we shall write $R^q_B\xi$ in place of $R^q_{T_B}\xi$ in keeping with P^q_B. Since $P^q_B U^q f = U^q f$ when f vanishes off B it follows from (4.2) and (4.7) that

$$\text{(4.13)} \qquad R^q_B\xi = \xi \quad \text{on} \quad B \quad \text{for} \quad \xi \in \text{Exc}^q,\ q \geq 0\,.$$

We are now going to describe the operators R^q_B in a, hopefully, familiar situation. We suppose that X and $\hat{X}$ are *standard* processes in strong duality relative to a σ-finite measure m as in [**BG**, §VI-1] with potential density kernel $u(x,y)$. Writing $(f,g) = \int fg\,dm$ we have the key switching identity for $B \in \mathcal{E}$—[**BG**, VI-(1.16)]—

$$\text{(4.14)} \qquad (P_B U f, g) = (f, \hat{P}_B \hat{U} g)\,.$$

For simplicity we shall suppose that X and $\hat{X}$ are transient. By [**BG**, VI-(1.11)] a measure ξ is excessive if and only if it has the form $\xi = \hat{h}m$ with $\hat{h} \in \hat{S}$ and $\hat{h} < \infty$ a.e. m.

(4.15) **Proposition**. *With X and $\hat{X}$ as above and $B \in \mathcal{E}$, if $\xi \in \text{Exc}$ with $\xi = \hat{h}m$ for $\hat{h} \in \hat{S}$, then $R_B\xi = (\hat{P}_B\hat{h})m$.*

Proof. Since $\hat{X}$ is transient there exist $\hat{U}f_k \uparrow \hat{h}$. Let $\mu_k = f_k m$. Then

$$\mu_k U(g) = (f_k, Ug) = (\hat{U}f_k, g) \uparrow (\hat{h}, g) = \xi(g)\,,$$

and so $\mu_k U \uparrow \xi$. But then by (4.8ii)

$$\begin{aligned} R_B\xi(g) &= \lim \mu_k P_B U g = \lim\,(f_k, P_B U g) \\ &= \lim\,(\hat{P}_B\hat{U}f_k, g) = (\hat{P}_B\hat{h}, g)\,. \end{aligned}$$

Consequently $R_B\xi = (\hat{P}_B\hat{h})m$. ∎

Remark. This result may be extended to the case where X and $\hat{X}$ are Borel right processes in weak duality relative to m as in [**GS84**]. In this case if $\xi = \hat{h}m$, then $R_B\xi = (\hat{P}_{B-}\hat{h})m$

where $\hat{P}_{B-} := \hat{P}_{\hat{S}_B}$ and $\hat{S}_B := \inf\{t > 0 : \hat{X}_{t-} \in B\}$. We leave the details to the interested reader.

There is one situation in which one may explicitly identify $R_T\xi$. If T is an exact terminal time, define

$$\phi(x) = P^x(T < \infty). \tag{4.16}$$

Clearly $\phi \in S$.

(4.17) **Proposition.** *Let* $\xi \in \text{Con}$ *and let* T *be an exact terminal time.* (i) $\phi = \phi^2$ *a.e.* ξ. (ii) $R_T\xi = \phi\xi$.

Proof. Using (2.16) we have for each $t > 0$ and ξ a.e. x

$$\begin{aligned}\phi(x) = P^x(T < \infty) &= P^x(T \le t) + P^x(t < T < \infty)\\ &= P^x(T \le t) + P^x(\phi \circ X_t;\, t < T)\\ &= P^x(T \le t) + \phi(x)P^x(t < T).\end{aligned}$$

Letting $t \to \infty$ this yields $\phi = \phi + \phi(1-\phi)$ or $\phi = \phi^2$ a.e. ξ, proving (i). For (ii) let us suppose for the moment only that $\xi \in \text{Inv}$. We write (Q_t) and (V^q) for the semigroup and resolvent of X killed at time T; that is,

$$Q_t f(x) = P^x(f \circ X_t;\, t < T), \tag{4.18}$$

$$V^q f(x) = P^x \int_0^T e^{-qt} f \circ X_t\, dt.$$

Then one has the identity for $q \ge 0$

$$U^q = V^q + P_T^q U^q. \tag{4.19}$$

Since $\xi \in \text{Inv}$, $\xi = q\xi U^q$ for $q > 0$ and so

$$R_T\xi = \uparrow \lim_{q\downarrow 0} R_T^q \xi = \uparrow \lim_{q\downarrow 0} q\xi P_T^q U^q. \tag{4.20}$$

Now $\xi Q_t \le \xi P_t \le \xi$—in fact ξQ_t is dominated by the restriction of ξ to $\{x : P^x(T > 0) = 1\}$. Therefore $\xi Q_t \downarrow \xi^i$ as

$t \to \infty$ where ξ^i is the invariant part of ξ relative to the semigroup (Q_t). By the usual argument we also have $q\xi V^q \downarrow \xi^i$ as $q \downarrow 0$. Combining these remarks with (4.19) and (4.20) we obtain

$$R_T\xi = \xi - \xi^i, \tag{4.21}$$

a formula valid for any $\xi \in \text{Inv}$. Next suppose $\xi \in \text{Con}$. Then $\phi = \phi^2$ a.e. ξ and, of course, $0 \le \phi \le 1$ everywhere. Let $\psi = 1 - \phi$ so that $\psi(x) = P^x\ (T = \infty)$. Clearly $\psi = \psi^2$ a.e. ξ. Let $f \in pb\mathcal{E}$ with $f > 0$ and $\xi(f) < \infty$. Write $Q_tf = Q_t(\psi f) + Q_t(\phi f)$. Using (2.16) we have

$$\begin{aligned} \xi Q_t(\psi f) &= P^\xi[\psi \circ X_t f \circ X_t;\ t < T] \\ &= P^\xi[\psi \circ X_0 f \circ X_t;\ t < T] \\ &= \int \xi(dx)\psi(x)Q_tf(x) \\ &= \xi(\psi Q_tf). \end{aligned} \tag{4.22}$$

If $\psi(x) = 1$, $T = \infty$ a.s. P^x and so $Q_tf(x) = P_tf(x)$. Consequently because ψ takes only the values zero and one

$$\xi(\psi Q_tf) = \xi(\psi P_tf) = \xi[(P_t(\psi f)] = \xi(\psi f)$$

where the second equality follows from a computation similar to (4.22). From (4.21), $\xi^i = \xi - R_T\xi \in \text{Inv}$ because $\xi \in \text{Con}$ implies $R_T\xi \in \text{Con}$. But then $\xi^i \in \text{Con}$ since $\xi^i \le \xi$. Therefore $P_t1 = 1$ a.e. ξ^i, and letting t approach infinity we obtain $P^{\xi^i}(\zeta < \infty) = 0$. Consequently

$$0 = \xi^iP_t(f) - \xi^iQ_t(f) = P^{\xi^i}[f \circ X_t;\ T \le t].$$

But $f > 0$ and so $P^{\xi_i}(T \le t) = 0$. Letting $t \to \infty$ we obtain $\xi^i(\phi) = P^{\xi^i}(T < \infty) = 0$. Hence

$$\xi Q_t(\phi f) \longrightarrow \xi^i(\phi f) = 0 \tag{4.23}$$

as $t \to \infty$. Combining these results we see that $\xi Q_t(f) \to \xi(\psi f)$ as $t \to \infty$; that is, $\xi^i = \psi\xi$. Thus (4.17ii) follows from (4.21). ∎

(4.24) **Remark.** I had originally proved (4.17ii) only under the additional assumption that ξ is finite. The extension to general $\xi \in$ Con is due to J. Steffens (unpublished). For an alternate proof using Kuznetsov measures, see [**GSt87**, (6.12)].

We now look at the reduction of an excessive measure ξ on a set $B \in \mathcal{E}^*$ defined by

$$(4.25) \qquad R_B^*\xi := \inf\{\eta \in \text{Exc}: \eta \geq \xi \quad \text{on} \quad B\}$$

for $\xi \in$ Exc and $B \in \mathcal{E}^*$. Of course, $\eta \geq \xi$ on B means $\eta(A) \geq \xi(A)$ for all $A \subset B$, $A \in \mathcal{E}^*$, or equivalently, $\eta \geq 1_B\xi$. Since the minimum of two excessive measures is again excessive one may as well restrict oneself to those η which are dominated by ξ in taking the infimum in (4.25). It is then clear that the infimum exists in the lattice of σ-finite measures and that $R_B^*\xi \in$ Exc with $R_B^*\xi \leq \xi$. We are going to identify $R_B^*\xi$ following Fitzsimmons [**F88b**].

Fix $B \in \mathcal{E}^*$ and let σ_B be the *Lebesgue penetration time* of B as defined in Appendix B, (B.10). Thus $\sigma_B := \inf\{t: \ell^*(\{s \leq t: X_s \in B\}) > 0\}$ where ℓ^* is Lebesgue outer measure. According to (B.10), σ_B is an exact terminal which is even an $(\mathcal{F}_{t+}^*)$ stopping time. In addition, a.s. the function $s \to 1_B \circ X_s$ is Lebesgue measurable and

$$\sigma_B = \inf\{t: \int_0^t 1_B \circ X_s \, ds > 0\}.$$

If $B^* := \{x: P^x(\sigma_B = 0) = 1\} \in \mathcal{E}^e$, then $\sigma_B = T_{B^*}$ a.s. All of these facts about σ_B are contained in (B.10). We may now describe $R_B^*\xi$.

(4.26) **Theorem.** (Fitzsimmons). *For $\xi \in$ Exc and $B \in \mathcal{E}^*$ one has $R_B^*\xi = R_{\sigma_B}\xi = R_{B^*}\xi$.*

Proof. Since $\sigma_B = T_{B^*}$ a.s., it is clear that $R_{\sigma_B}\xi = R_{B^*}\xi$. Because $\int_0^{\sigma_B} 1_B \circ X_s\, ds = 0$, it follows that if f vanishes off B, then

$$\int_{\sigma_B}^{\infty} e^{-qt} f \circ X_t\, dt = \int_0^{\infty} e^{-qt} f \circ X_t\, dt,$$

and, hence, $P^q_{\sigma_B} U^q f = U^q f$. Therefore for such an f and $q > 0$

$$R^q_{\sigma_B}\xi(f) = L^q(\xi, P^q_{\sigma_B} U^q f) = L^q(\xi, U^q f) = \xi(f),$$

and letting $q \downarrow 0$ we obtain $R_{\sigma_B}\xi = \xi$ on B. Consequently $R^*_B\xi \le R_{\sigma_B}\xi$. For the reverse inequality we require a well-known fact that is due to Hunt. Although this appears as exercise II-(2.20) in **[BG]** we shall state it as a lemma and give the proof.

(4.27) **Lemma.** *Let $q > 0$ and $g, h \in bp\mathcal{E}^*$. Define*

(4.28) $\psi(x) =$

$$P^x \int_0^{\infty} e^{-qt} g \circ X_t \left[1 - \exp\left(-\int_0^t h \circ X_s\, ds\right)\right] dt.$$

If $\varphi = U^q g$, then $\psi \le \varphi$ and $\psi = U^q[h(\varphi - \psi)]$.

Proof. Clearly $\psi \le \varphi$. Using the identity

$$1 - \exp\left(-\int_0^t h \circ X_s\, ds\right)$$

$$= \int_0^t h \circ X_s \exp\left(-\int_0^s h \circ X_u\, du\right) ds,$$

we obtain on interchanging the order of integration

$$\psi(x) = P^x \int_0^{\infty} e^{-qs} \varphi \circ X_s h \circ X_s \exp\left(-\int_0^s h \circ X_u\, du\right) ds.$$

Next substituting the identity

$$1 - \exp\left(-\int_0^s h \circ X_u \, du\right)$$

$$= \int_0^s h \circ X_t \exp\left(-\int_t^s h \circ X_u \, du\right) dt$$

into this last expression for ψ, changing the order of integration, and using the Markov property yields $\psi = U^q[h(\varphi - \psi)]$. ∎

We now return to the proof of (4.26). Let $h_n = n1_B$. Given $g \in pb\mathcal{E}^*$ and $q > 0$, let $\varphi = U^q g$. Define ψ_n by (4.28) with h replaced by h_n so that $\psi_n = U^q[h_n(\varphi - \psi_n)]$. But $\exp\left(-n \int_0^t 1_B \circ X_s \, ds\right) \downarrow 1_{[0,\sigma_B]}(t)$ as $n \to \infty$, and so from (4.28), $\psi_n \uparrow P^q_{\sigma_B} U^q g$. That is, there exist bounded functions f_n vanishing off B with $U^q f_n \uparrow P^q_{\sigma_B} U^q g$. Therefore if $\eta \in \text{Exc}$ with $\eta \geq \xi$ on B, writing $\sigma = \sigma_B$ we have

$$\begin{aligned}
\eta(g) &= L^q(\eta, U^q g) \geq L^q(\eta, P^q_\sigma U^q g) \\
&= \lim L^q(\eta, U^q f_n) = \lim \eta(f_n) \geq \lim \xi(f_n) \\
&= \lim L^q(\xi, U^q f_n) = L^q(\xi, P^q_\sigma U^q g) = R^q_\sigma \xi(g) .
\end{aligned}$$

Hence $\eta \geq R^q_\sigma \xi$. Letting $q \downarrow 0$ gives $\eta \geq R_\sigma \xi$ and consequently $R^*_B \xi \geq R_{\sigma_B} \xi$. ∎

(4.29) **Remark.** If $\sigma_B = T_B$ a.s., for example if B is finely open, then $R_B = R^*_B$. This result for B finely open goes back to Hunt [**H57**].

We close this section with the following important relationship between R^q_T and P^q_T and some of its consequences.

(4.30) **Theorem.** *Let T be an exact terminal time. If $0 \leq r < q$ and $\xi \in \text{Exc}^r$, then $(R^q_T \xi) P^r_T = (R^r_T \xi) P^q_T$.*

Proof. We need several preliminary identities for the proof. Define for $q \geq 0$ and $f \in p\mathcal{E}^*$

$$V^q f(x) := P^x \int_o^T e^{-qt} f \circ X_t \, dt .$$

Then $(V^q)_{q \geq 0}$ is the resolvent of X killed at time T. See [**S**, (12.16)]. It is immediate that for $q \geq 0$

$$U^q = V^q + P_T^q U^q . \tag{4.31}$$

In addition, since $T \circ \theta_t = T - t$ on $\{t < T\}$, for $r, q \geq 0$

$$\begin{aligned}
(4.32)\quad & (q-r) V^r P_T^q f \\
&= (q-r) P^{\cdot} \left[\int_0^T e^{-rt} e^{-qT \circ \theta_t} f \circ X_{t+T \circ \theta_t} \, dt ;\ T < \zeta \right] \\
&= P^{\cdot} [e^{-qT} f \circ X_T (e^{(q-r)T} - 1)] = P_T^r f - P_T^q f .
\end{aligned}$$

Write (4.31) with q replaced by r and let it act on P_T^q. Then use (4.32) to obtain

$$(q-r) U^r P_T^q + P_T^q = P_T^r + (q-r) P_T^r U^r P_T^q . \tag{4.33}$$

Suppose $0 < r < q$, then using (4.33) and (4.31)

$$\begin{aligned}
(4.34)\quad & [I + (q-r) U^r] P_T^q U^q \\
&= P_T^r [U^q + (q-r) U^r (U^q - V^q)] \\
&= P_T^r U^r - (q-r) P_T^r U^r V^q ,
\end{aligned}$$

where the last equality follows from the resolvent equation. Of course, the subtractions are justified since $0 < r < q$. Because of (4.32) with r and q interchanged, we obtain for $0 < r < q$

$$P_T^r U^r P_T^r - (q-r) P_T^r U^r V^q P_T^r = P_T^r U^r P_T^q .$$

Now applying (4.34) to P_T^r and using this last relationship we find

$$[I + (q-r)U^r] P_T^q U^q P_T^r = P_T^r U^r P_T^q , \tag{4.35}$$

first for $0 < r < q$, and then letting $r \downarrow 0$ for $r = 0$ as well since the convergence is monotone increasing.

We now turn to proof of (4.30) proper. Since as r decreases to zero $P_T^r \uparrow P_T$ and $R_T^r \xi \uparrow R_T \xi$, it suffices to prove (4.30) when $r > 0$. But for $r > 0$, $\mathrm{Exc}^r = \mathrm{Dis}^r$ so that there exists a sequence $(\mu_n U^r)$ increasing to ξ. Hence by (4.3), $\mu_n P_T^r U^r \uparrow R_T^r \xi$. Define $\nu_n := \mu_n + (q-r)\mu_n U^r$ so that $\nu_n U^q \uparrow \xi$ and $\nu_n P_T^q U^q \uparrow R_T^q \xi$. Consequently in view (4.35)

$$\begin{aligned} (R_T^q \xi) P_T^r &= \lim_n \mu_n [I + (q-r)U^r] P_T^q U^q P_T^r \\ &= \lim_n \mu_n P_T^r U^r P_T^q = (R_T^r \xi) P_T^q . \qquad \blacksquare \end{aligned}$$

Here are several consequences of (4.17) and (4.30).

(4.36) Proposition. *Let $\xi \in \mathrm{Con}$ and T be an exact terminal time. Then*

(i) $(R_T \xi) P_T^q = \xi P_T^q$ *for* $q \geq 0$,

(ii) $R_T^q \xi = q (R_T \xi) P_T^q U^q$ *for* $q > 0$.

Proof. Let ϕ be defined as in (4.16) so that $R_T \xi = \phi \xi$. Also a.e. ξ, $\phi(x)$ is either zero or one, and if $\phi(x) = 0$ then $P_T^q(x, \cdot) = 0$ for all $q \geq 0$. Therefore $(R_T \xi) P_T^q = (\phi \xi) P_T^q = \xi P_T^q$ proving (i). For (ii) since $\xi = q \xi U^q$ if $q > 0$, $R_T^q \xi = q \xi P_T^q U^q = q (R_T \xi) P_T^q U^q$ because of (i). $\blacksquare$

The following result should be compared with (3.8).

(4.37) Proposition. *Let* $0 \le r < q$, $\xi \in \text{Exc}^r$, $u \in S^r$, *and* T *be an exact terminal time. Then*

$$\begin{aligned} L^q(\xi, P_T^q u) &= L^r(\xi, P_T^r u) + (q-r) R_T^r \xi(P_T^q u) \\ &= L^r(\xi, P_T^r u) + (q-r) R_T^q \xi(P_T^r u)\,. \end{aligned}$$

Proof. Replacing X by X^r—the r-subprocess of X—it suffices to prove this when $r = 0$. The second equality follows from (4.30). For the first suppose firstly that $\xi \in \text{Con}$. Then $L(\xi, P_T u) = 0$ and $\xi = q\xi U^q$. Therefore

$$L^q(\xi, P_T^q u) = q\xi(P_T^q u) = qR_T\xi(P_T^q u)$$

because of (4.36i). If $\xi \in \text{Dis}$, then there exist $\mu_n U \uparrow \xi$ and $\mu_n P_T U \uparrow R_T\xi$. Define $\nu_n := \mu_n(I + qU)$ so that $\nu_n U^q \uparrow \xi$. Now using (4.33) for the second equality,

$$\begin{aligned} L^q(\xi, P_T^q u) &= \lim_n \mu_n[P_T^q + qUP_T^q](u) \\ &= \lim_n \mu_n[P_T + qP_T U P_T^q](u) \\ &= L(\xi, P_T u) + qR_T\xi(P_T^q u)\,. \end{aligned}$$ ∎

Remark. The operator R_T^q on Exc^q is often called Hunt's *balayage* operation on excessive measures in the literature. See [**Hu57**].

5. Potential Theory of Excessive Measures

In this section we shall develop some of the basic potential theory of the cone of excessive measures. We shall then prove a preliminary version of an important integral representation theorem of Fitzsimmons [**F88b**]. Additional properties of the cone Exc will appear as corollaries to this integral representation. The definitive form of Fitzsimmons' theorem appears in (7.10). Its statement requires the use of Kuznetsov measures to be developed in §6. Theorems 5.9 and 5.11 below are more or less well-known. The analogous results for excessive functions may be found in [**DM**, XII], for example. However, proofs of the results we need are somewhat scattered in the literature and so we shall give a systematic development. Our development follows [**F88b**].

In this section only we shall denote by Dex the class of σ-finite measures λ on E for which there exists $\xi \in \text{Exc}$ with $\lambda \le \xi$. The reduction of $\lambda \in \text{Dex}$ is defined by

$$(5.1) \qquad R\lambda := \inf\{\xi \in \text{Exc}: \xi \ge \lambda\},$$

where infimum is taken in the lattice of σ-finite measures. Clearly $R\lambda \in \text{Exc}$ and is the smallest excessive measure dominating λ. It is obvious that $R(\alpha\lambda) = \alpha R\lambda$ if $0 \le \alpha < \infty$, $R\lambda \le R\mu$ if $\lambda \le \mu \in \text{Dex}$, and $R(\lambda + \mu) \le R\lambda + R\mu$ if $\lambda, \mu \in \text{Dex}$. In the remainder of this section all named measures are in Dex and all named sets are in $\mathcal{E}^*$. The reduction of λ on $A \subset E$ is defined by

$$(5.2) \qquad R^*_A\lambda := R(1_A \cdot \lambda).$$

If $\xi \in \text{Exc}$, then $R^*_A\xi$ as defined in (5.2) agrees with the definition of $R^*_A\xi$ in (4.25). It is evident that if $A \subset B$, then

$R^*_A\lambda \le R^*_B\lambda$. By (4.26), if $\xi \in \text{Exc}$, then $R^*_A\xi = R_{\sigma_A}\xi$. Consequently R^*_A is *additive* on Exc; that is,

(5.3) $$R^*_A(\eta+\xi) = R^*_A\eta + R^*_A\xi\,; \quad \eta,\xi \in \text{Exc}\,.$$

If $\lambda_n \uparrow \lambda$, then $\xi := \uparrow \lim R\lambda_n \le R\lambda$ and $\xi \in \text{Exc}$. But $\xi \ge \lambda_n$ for all n and so $\xi \ge \lambda$. As a result

(5.4) $$\lambda_n \uparrow \lambda \Longrightarrow R\lambda_n \uparrow R\lambda\,.$$

If λ and μ are σ-finite measures and γ is any σ-finite measure dominating both λ and μ (for example $\lambda+\mu$ or $\lambda \vee \mu$), define

(5.5) $$\{\lambda \le \mu\} = \left\{x\colon \frac{d\lambda}{d\gamma}(x) \le \frac{d\mu}{d\gamma}(x)\right\}.$$

It is easy to check that $\{\lambda \le \mu\}$ is uniquely determined $\text{mod}(\lambda \vee \mu)$ independently of the choice of γ and we may and do suppose $\{\lambda \le \mu\} \in \mathcal{E}$. Note that $\text{mod}(\lambda\vee\mu)$, $B \subset \{\lambda \le \mu\}$ if and only if $\lambda \le \mu$ on B. Of course, $\{\lambda < \mu\} \in \mathcal{E}$ is defined in the same manner. The next result is due to Mokobodski. See [**He74**].

(5.6) **Proposition.** *Let* $\xi = R\lambda$. *If* $0 < \epsilon < 1$ *and* $\{\epsilon\xi < \lambda\} \subset A$, *then* $\xi = R^*_A\xi$.

Proof. Let $\eta = R^*_A\xi \in \text{Exc}$. Then $1_A\xi \le \eta \le \xi$, and so $\eta = \xi$ on A. On $A^c \subset \{\epsilon\xi \ge \lambda\}$ one has $\lambda \le \epsilon\xi \le \epsilon\xi + (1-\epsilon)\eta$ while on A, $\lambda \le R\lambda = \xi = \epsilon\xi + (1-\epsilon)\eta$ since $\eta = \xi$ on A. Therefore $\lambda \le \epsilon\xi + (1-\epsilon)\eta$ everywhere. Hence $\xi = R\lambda \le \epsilon\xi + (1-\epsilon)\eta$ and consequently $\xi \le \eta$. Thus $\eta = \xi$. ∎

For any two σ-finite measures λ and μ we define $(\lambda - \mu)_+ = (\lambda \vee \mu) - \mu$. It is easy to check that $(\lambda-\mu)_+ \le \lambda$ and that if λ, μ, and ν are σ-finite, then

(5.7) (i) $$\nu \ge (\lambda-\mu)_+ \Longleftrightarrow \nu + \mu \ge \lambda\,,$$

(ii) $$\{\nu < (\lambda-\mu)_+\} = \{\nu + \mu < \lambda\}\,.$$

We now define $R(\lambda-\mu)=R[(\lambda-\mu)_+]$ for $\lambda\in\mathrm{Dex}$. Then using (5.7i)

$$(5.8)\qquad \begin{aligned} R(\lambda-\mu) &= \inf\{\eta\in\mathrm{Exc}\colon \eta\geq(\lambda-\mu)_+\}\\ &= \inf\{\eta\in\mathrm{Exc}\colon \eta+\mu\geq\lambda\}. \end{aligned}$$

Of course $R(\lambda-\mu)\in\mathrm{Exc}$ and $\mu+R(\lambda-\mu)\geq\lambda$.

(5.9) **Theorem.** *Let* $\xi, m_1, m_2\in\mathrm{Exc}$ *with* $\xi\leq m_1+m_2$. *Then there exist* $\xi_1,\xi_2\in\mathrm{Exc}$ *with* $\xi_1\leq m_1$, $\xi_2\leq m_2$, *and* $\xi=\xi_1+\xi_2$.

Proof. Define $\xi_2=R(\xi-m_1)$ and $\xi_1=R(\xi-\xi_2)$. Then $\xi_1,\xi_2\in\mathrm{Exc}$ and since $m_1+m_2\geq\xi$, one has $\xi_2\leq m_2$. Also $\xi\leq m_1+R(\xi-m_1)=m_1+\xi_2$, and so $\xi_1\leq m_1$. Since $\xi\leq\xi_2+R(\xi-\xi_2)=\xi_2+\xi_1$, it remains to show $\xi_1+\xi_2\leq\xi$. To this end, given $0<\epsilon<1$, define $A=\{\epsilon(\xi_1+\xi_2)\leq\xi\}$. Then $A_1\colon=\{\epsilon\xi_1+\xi_2<\xi\}\subset A$ and by (5.7ii), $A_1=\{\epsilon\xi_1<(\xi-\xi_2)_+\}$. Using (5.6) we obtain $\xi_1=R_A^*\xi_1$. Next define $A_2=\{\epsilon\xi_2+m_1<\xi\}$. Because $\xi_1\leq m_1$ we have $A_2\subset\{\epsilon(\xi_2+m_1)<\xi\}\subset A$, and using (5.6) again gives $\xi_2=R_A^*\xi_2$. Now in light of (5.3)

$$\begin{aligned} \xi_1+\xi_2 &= R_A^*\xi_1+R_A^*\xi_2=R_A^*(\xi_1+\xi_2)\\ &= \epsilon^{-1}R_A^*[\epsilon(\xi_1+\xi_2)]\leq\epsilon^{-1}\xi, \end{aligned}$$

where the last equality follows because $\epsilon(\xi_1+\xi_2)\leq\xi$ on A. Letting $\epsilon\uparrow 1$ gives $\xi_1+\xi_2\leq\xi$. ■

The *strong* or *specific order* "$\prec$" on Exc is defined for $\xi,\eta\in\mathrm{Exc}$ by

$$(5.10)\quad \eta\prec\xi \iff \text{there exists}\quad \lambda\in\mathrm{Exc}\quad\text{with}\quad \eta+\lambda=\xi.$$

Of course, this is equivalent to $\eta\leq\xi$ and $\xi-\eta\in\mathrm{Exc}$. We refer to $\eta\leq\xi$, that is, $\xi(A)\leq\eta(A)$ for all $A\in\mathcal{E}$, as the *simple* or *natural order* on Exc.

(5.11) **Theorem.** *Let $\xi, m \in$ Exc and define $\rho = R(\xi - m)$. Then there exists a unique $\lambda \in$ Exc with $\rho + \lambda = \xi$ and $\lambda \leq m$. In fact $\lambda = \sup\{\eta \in \text{Exc}: \eta \leq m,\ \eta \prec \xi\}$.*

Proof. Since $\xi \leq \rho + m$ we may write $\xi = \xi_1 + \xi_2$ with $\xi_1, \xi_2 \in$ Exc, $\xi_1 \leq \rho$ and $\xi_2 \leq m$ by (5.9). Then $\xi_1 + m \geq \xi_1 + \xi_2 = \xi$ and so $\xi_1 \geq R(\xi - m) = \rho$. Hence $\xi_1 = \rho$. Set $\lambda = \xi_2$ so $\xi = \rho + \lambda$ and $\lambda \leq m$. For the last assertion suppose $\eta \in$ Exc, $\eta \leq m$, and $\eta + \gamma = \xi$ for some $\gamma \in$ Exc. Then $\gamma + m \geq \gamma + \eta = \xi$. Thus $\gamma \geq \rho$. But $\gamma + \eta = \xi = \rho + \lambda$ and since $\rho \leq \gamma$ one must have $\eta \leq \lambda$. ∎

It follows from Theorem 5.9 that Exc is an H-cone in the terminology of [**BBC81**]. The next two results are standard facts about H-cones. See [**BBC81**]. However, we shall include proofs for completeness.

(5.12) **Proposition.** Exc *is a lattice in the strong order.*

Proof. Given ξ_1 and ξ_2 in Exc we must show the existence of the least upper bound and greatest lower bound of ξ_1 and ξ_2 in the strong order. These will be denoted by $\xi_1 \curlyvee \xi_2$ and $\xi_1 \curlywedge \xi_2$ respectively while "$\vee$" and "$\wedge$" will always refer to the simple order of measures. Let $H = \{\eta \in \text{Exc}: \xi_1 \prec \eta,\ \xi_2 \prec \eta\}$ and set $\eta_0 = \inf H$. Of course, this is the infimum in the simple order of measures which exists as a σ-finite measure because $\xi_1 + \xi_2 \in H$. Clearly $\eta_0 \in$ Exc. If $\eta \in H$, then $\eta - \xi_i \in$ Exc for $i = 1, 2$. It follows that

$$\eta_0 - \xi_i = \inf\{\eta - \xi_i : \eta \in H\} \in \text{Exc};\ i = 1, 2,$$

and so $\eta_0 \in H$. To show that $\eta_0 = \xi_1 \curlyvee \xi_2$ it remains to show that if $\eta \in H$, then $\eta_0 \prec \eta$. To this end fix $\eta \in H$ and let $\rho = R(\eta - \eta_0)$. By (5.11) there exists $\lambda \in$ Exc with $\lambda + \rho = \eta$ and $\lambda \leq \eta_0$. But $\rho = R[(\eta - \xi_i) - (\eta_0 - \xi_i)]$, $i = 1, 2$, and hence there exist $\lambda_i \in$ Exc with $\lambda_i + \rho = \eta - \xi_i$ and $\lambda_i \leq \eta_0 - \xi_i$, $i = 1, 2$. Therefore $\rho \prec \eta - \xi_i$ and so there exist $\mu_i \in$ Exc with $\eta = \rho + \mu_i + \xi_i$. Since $\eta = \lambda + \rho$ we conclude that $\lambda \in H$ and hence, $\eta_0 \leq \lambda$. Therefore $\lambda = \eta_0$ and so $\eta_0 \prec \eta$. Consequently

$\eta_0 = \xi_1 \curlyvee \xi_2$. Because $\xi_1 + \xi_2$ strongly dominates ξ_1 and ξ_2, $\xi := \xi_1 + \xi_2 - (\xi_1 \curlyvee \xi_2) \in \mathrm{Exc}$ and it is easily checked that $\xi = \xi_1 \curlywedge \xi_2$. ∎

Remarks. The same argument shows that Exc is a conditionally complete lattice in the strong order. See [**BBC81**, Th. 2.1.5]. In (5.28) we shall give a more explicit recipe for $\xi_1 \curlyvee \xi_2$ and $\xi_1 \curlywedge \xi_2$.

The next result is a useful refinement of (5.9).

(5.13) **Corollary.** *Let $\xi_1, \xi_2, m_1, m_2 \in \mathrm{Exc}$ with $\xi_1 + \xi_2 = m_1 + m_2$. Then there exist $\eta_{ij} \in \mathrm{Exc}$, $1 \le i, j \le 2$ such that $\xi_i = \eta_{i1} + \eta_{i2}$, $i = 1, 2$ and $m_j = \eta_{1j} + \eta_{2j}$, $j = 1, 2$.*

Proof. Define $\eta_{11} = m_1 \curlywedge \xi_1$, $\eta_{12} = \xi_1 - \eta_{11}$, and $\eta_{21} = m_1 - \eta_{11}$. Then $\xi_1 = \eta_{11} + \eta_{12}$ and $m_1 = \eta_{11} + \eta_{21}$. Because $\xi_1 + \xi_2 = m_1 + m_2$, $\eta_{22} := \xi_1 + \xi_2 - m_1 \curlyvee \xi_1 \in \mathrm{Exc}$. Using $m_1 \curlywedge \xi_1 + m_1 \curlyvee \xi_1 = \xi_1 + m_1$ and $\xi_1 + \xi_2 = m_1 + m_2$, one readily checks that $\xi_2 = \eta_{21} + \eta_{22}$ and $m_2 = \eta_{12} + \eta_{22}$. ∎

We now turn to the development of the preliminary version of Fitzsimmons' theorem referred to earlier. Fix $\xi, m \in \mathrm{Exc}$. Define

$$\gamma_t = R(\xi - tm), \quad t \ge 0. \tag{5.14}$$

Clearly $\gamma_0 = \xi$, $\gamma_t \le \gamma_s$ if $s \le t$. Hence we may define $\gamma_\infty = \downarrow \lim_{t\to\infty} \gamma_t$ and $\gamma_\infty \in \mathrm{Exc}$. Let $\Gamma_t = (\xi - tm)_+$ so that $\gamma_t = R\Gamma_t$, $t \ge 0$. Let $\lambda = \xi + m$. If $f \ge 0$ with $\lambda(f) < \infty$, then

$$\Gamma_t(f) = \int \left(\frac{d\xi}{d\lambda} \vee t\,\frac{dm}{d\lambda} - t\,\frac{dm}{d\lambda} \right) f\,d\lambda .$$

For each fixed x the integrand above is a continuous decreasing convex function of t on $[0, \infty[$, and hence so is $t \to \Gamma_t(f)$. If $\alpha, \beta \ge 0$ with $\alpha + \beta = 1$ and $t, s \ge 0$, then

$$\gamma_{\alpha t + \beta s} = R(\Gamma_{\alpha t + \beta s}) \le R(\alpha \Gamma_t + \beta \Gamma_s) \le \alpha \gamma_t + \beta \gamma_s .$$

Consequently $t \to \gamma_t(f)$ is a decreasing convex function on $[0,\infty[$ if $f \geq 0$ with $\gamma_0(f) = \xi(f) < \infty$. In particular $t \to \gamma_t(f)$ is continuous on $]0,\infty[$. Since $\Gamma_t \uparrow \Gamma_0 = \xi$ as $t \downarrow 0$ it follows from (5.4) that $\gamma_t \uparrow \xi = \gamma_0$ as $t \downarrow 0$ so $t \to \gamma_t(f)$ is, in fact, continuous on $[0,\infty[$.

Next we want to choose a "good" version of $d\gamma_t/d\lambda$. Let g_t^0 be any Borel measurable version of this Radon-Nikodym derivative that satisfies $0 \leq g_t^0(x) \leq 1$ for all t and x. We may suppose that if α, s, t are rational with $\alpha \in [0,1]$, $0 \leq s < t$, and $\beta = 1 - \alpha$, then everywhere on E, $g_t^0 \leq g_s^0$ and $g_{\alpha s+\beta t}^0 \leq \alpha g_s^0 + \beta g_t^0$. Define for $t \geq 0$

$$g_t(x) = \sup_{\substack{r>t \\ r\in\mathbb{Q}}} g_r^0(x) =\uparrow \lim_{\substack{r\downarrow t \\ r\in\mathbb{Q}}} g_r^0(x)\,.$$

Then for each x, $t \to g_t(x)$ is a decreasing convex function on $[0,\infty[$ that is continuous even at $t = 0$. Because $t \to \gamma_t(f)$ is continuous when $f \geq 0$ with $\gamma_0(f) < \infty$, it follows that for each $t \geq 0$, $g_t = d\gamma_t/d\lambda$. In the sequel g_t will always denote this version of $d\gamma_t/d\lambda$. Since $\gamma_0 = \xi$, $g_0 = d\xi/d\lambda$.

Let b be a fixed Borel measurable version of $dm/d\lambda$ with $0 \leq b \leq 1$. Define for $t \geq 0$ and $\epsilon > 0$

$$\text{(5.15)} \qquad A(t,\epsilon) = \{(1+\epsilon t)g_0 \geq tb + g_t\}\,.$$

Each $A(t,\epsilon) \in \mathcal{E}$ and clearly $A(t,\epsilon)$ increases with ϵ. Since $t \to g_t(x)$ is convex, if the linear function $t \to g_0(x)+(\epsilon g_0(x)-b(x))t$ is greater than or equal to $g_t(x)$ for $t = t_0$, then the same inequality holds for $t \leq t_0$. As a result, $A(t,\epsilon) \subset A(s,\epsilon)$ if $0 \leq s \leq t$ and $\epsilon > 0$. Using (5.7ii) and $g_t \leq g_0$ we have if $0 < \epsilon t < 1$,

$$A(t,\epsilon) \supset \{g_0 \geq tb + (1-\epsilon t)g_t\} = \{(\xi - tm)_+ \geq (1-\epsilon t)\gamma_t\}\,.$$

Therefore if $0 < \epsilon t < 1$ it follows from (5.6) that $\gamma_t = R^*_{A(s,\epsilon)}\gamma_t$ for $0 \leq s \leq t$, and since $A(s,\epsilon)$ increases with ϵ, it follows that

$$\text{(5.16)} \qquad \gamma_t = R^*_{A(s,\epsilon)}\gamma_t\,; \quad 0 \leq s \leq t,\ \ \epsilon > 0\,.$$

Define

$$(5.17) \qquad \delta_t = \downarrow \lim_{\epsilon \downarrow 0} R^*_{A(t,\epsilon)} m = \downarrow \lim_{\epsilon \downarrow 0} R_{\sigma_{A(t,\epsilon)}} m, \quad t \geq 0;$$

$$(5.18) \quad T(t) = \uparrow \lim_{\epsilon \downarrow 0} \sigma_{A(t,\epsilon)}, \quad t \geq 0,$$

where, of course, σ_A denotes the Lebesgue penetration time of A as in §4. For each t, $\delta_t \in$ Exc and $T(t)$ is a terminal time as defined at the beginning of §4. Moreover each $T(t)$ is an $(\mathcal{F}^*_{t+})$ stopping time. However, $T(t)$ need not be exact. Clearly $0 \leq s < t$ implies $\delta_t \leq \delta_s$ and $T(s) \leq T(t)$.

(5.19) **Lemma.** *If* $0 \leq s < t$, *then* $\gamma_t + (t-s)\delta_t \leq \gamma_s \leq \gamma_t + (t-s)\delta_s$.

Proof. If $\eta \in$ Exc and $\eta + tm \geq \xi$, then $\gamma_s = R(\xi - sm) \leq \eta + (t-s)m$, and hence $\gamma_s \leq \gamma_t + (t-s)m$. Using (5.16) we obtain $\gamma_s \leq \gamma_t + (t-s)R^*_{A(s,\epsilon)}m$. Letting $\epsilon \downarrow 0$ yields the inequality on the right hand side of (5.19). For the other side note that since $\gamma_s + sm \geq \xi$ one has $\gamma_s + sm + \epsilon t\xi \geq (1+\epsilon t)\xi$ and on $A(t,\epsilon)$, $(1+\epsilon t)\xi \geq tm + \gamma_t$. Hence on $A(t,\epsilon)$, $\gamma_s + \epsilon t \xi \geq \gamma_t + (t-s)m$. Therefore using (5.3) and (5.16)

$$\gamma_s + \epsilon t \xi \geq R^*_{A(t,\epsilon)}(\gamma_t + (t-s)m) = \gamma_t + (t-s)R^*_{A(t,\epsilon)}m,$$

and letting $\epsilon \downarrow 0$ we obtain the left hand side of (5.19). ∎

It is immediate from (5.19) that

$$\delta_t \leq \frac{\gamma_s - \gamma_t}{t-s} \leq \delta_s, \quad 0 \leq s < t.$$

If $f \geq 0$ with $\lambda(f) < \infty$, letting $t \downarrow s$ the above gives $\delta_{s+}(f) \leq -D^+\gamma_s(f) \leq \delta_s(f)$ where D^+ denotes the right hand derivative. Since $s \to \delta_s(f)$ is decreasing there must be equality except for countably many s and since $t \to \gamma_t(f)$ is convex and continuous on $[0,\infty[$ one has (recall $\gamma_0 = \xi$)

$$(5.20) \qquad \xi(f) = \gamma_t(f) + \int_0^t \delta_u(f)\,du; \quad t \geq 0.$$

Letting $t \uparrow \infty$ and noting that if $\xi \le am$ for some $a < \infty$, then $\gamma_t = 0$ for all $t \ge a$ establishes the following result.

(5.21) **Theorem.** (Fitzsimmons) *Let ξ and m be excessive measures. Then the decreasing family $(\delta_t)_{t\ge 0}$ of excessive measures defined by* (5.17) *satisfies*

$$\xi = \gamma_\infty + \int_0^\infty \delta_t \, dt \,, \tag{5.22}$$

where $\gamma_\infty = \downarrow \lim_{t\to\infty} R(\xi - tm)$. If, in addition, for some $a < \infty$, $\xi \le am$, then $\xi = \int_0^a \delta_t \, dt$.

Suppose that m is a potential, $m = \mu U$. Then if (T_n) is an increasing sequence of exact terminal times with limit T one has $R_{T_n}(\mu U) = \mu P_{T_n} U \uparrow \mu P_T U$. Since $R^*_{A(t,\epsilon)} m = R_{\sigma_{A(t,\epsilon)}} m$ by (4.26), it follows from (5.17) and (5.18) that $\delta_t = \mu P_{T(t)} U$ in this case. Therefore when $\xi \le \mu U$ we have

$$\xi = \int_0^1 \mu P_{T(t)} U \, dt = \left(\int_0^1 \mu P_{T(t)} \, dt \right) U \,.$$

Thus we have established the following:

(5.23) **Corollary.** *Let $\xi \in$ Exc with $\xi \le \mu U \in$ Pot. Then $\xi = \nu U$ where $\nu = \int_0^1 \mu P_{T(t)} \, dt$ with $(T(t))_{0\le t\le 1}$ being the increasing family of terminal times given by* (5.18).

Remarks. If $f \ge 0$ with $\xi(f) < \infty$, then $t \to \delta_t(f)$ is decreasing and so $t \to \delta_t(f)$ is Borel measurable for all $f \in p\mathcal{E}$. In particular in (5.23), the family $(\mu P_{T(t)})$ is Borel measurable in t. Of course, (5.23) contains the Skorokhod embedding theorem for transient right processes which, under stronger hypotheses, is due to Rost. See [**Ro71**].

We can now establish an integral representation for purely excessive measures. Once again we follow Fitzsimmons, but this time [**F87b**]. We begin with a definition.

(5.24) **Definition.** A family $(\nu_t)_{t>0}$ of σ-finite measures on E is an *entrance law* (for (P_t)) provided $\nu_t P_s = \nu_{t+s}$ for $t, s > 0$.

If $f \in p\mathcal{E}$, $(s,x) \to P_s f(x)$ is $\mathcal{B}(]0,\infty[) \times \mathcal{E}^*$ measurable and hence it follows from the defining property of an entrance law that $t \to \nu_t(f)$ is $\mathcal{B}(]0,\infty[)$ measurable. Thus any entrance law (ν_t) is Borel measurable in t. It is immediate that if (ν_t) is an entrance law and $\xi := \int_0^\infty \nu_t \, dt$ is σ-finite, then ξ is purely excessive. The next result is the converse of this statement.

(5.25) **Theorem.** *Let $\xi \in$ Pur. Then there exists a unique entrance law (ν_t) such that $\xi = \int\limits_0^\infty \nu_t \, dt$.*

Proof. Fix $0 < r < s$ and set $\mu := (s-r)^{-1}[\xi P_r - P_s \xi]$. Then μ is a well-defined σ-finite measure because $\xi P_s \le \xi P_r$. Let $f \ge 0$ with $\xi(f) < \infty$. A routine calculation shows that for $a < \infty$

$$\int_0^a \mu P_t(f) \, dt = (s-r)^{-1} \left[\int_r^s \xi P_t(f) \, dt - \int_{r+a}^{s+a} \xi P_t(f) \, dt \right] .$$

Since $\xi \in$ Pur, $\xi P_t(f)$ decreases to zero as $t \to \infty$ and so letting $a \to \infty$ we obtain

$$\mu U(f) = (s-r)^{-1} \int_r^s \xi P_t(f) \, dt .$$

Hence $\mu U \le \xi$ and so μU is σ-finite. Also $\mu U \ge \xi P_s$ and since $\xi P_s \in$ Exc, it follows from (5.23) that $\xi P_s = \nu_s U$. The σ-finiteness of $\nu_s U$ implies that of ν_s. Now if $s, t > 0$

$$\nu_{t+s} U = \xi P_{t+s} = (\xi P_s) P_t = \nu_s U P_t = \nu_s P_t U ,$$

and hence $\nu_{t+s} = \nu_s P_t$ by (2.12). Therefore $(\nu_t)_{t>0}$ is an entrance law. But

$$\xi P_s = \nu_s U = \int_0^\infty \nu_s P_t \, dt = \int_s^\infty \nu_t \, dt ,$$

and letting $s \downarrow 0$ we obtain the representation in (5.25). It remains to prove the uniqueness. If $\xi = \int_0^\infty \mu_t \, dt$ where μ_t is another entrance law, than a simple calculation gives $\xi P_s = \mu_s U$ for $s > 0$. Since $\xi P_s = \nu_s U$, another appeal to (2.12) gives $\nu_s = \mu_s$. ∎

Remarks. Given Theorem 5.23 this is the simplest proof of (5.25) known to me. Other proofs may be found in [**Dy80**], [**FM86**], or [**GG87**]. A very general form of (5.25) under minimal hypotheses is given in [**J87**].

In case $\xi << m$ an alternate description of γ_∞ in (5.22) is given in [**F88b**].

(5.26) **Proposition.** *Let* ξ, $m \in \text{Exc}$ *with* $\xi << m$. *Let* ψ *be any Borel measurable version of* $d\xi/dm$. *Then in the representation* (5.22), $\gamma_\infty = \downarrow \lim_{t\to\infty} R^*_{\{\psi>t\}} \xi$.

Proof. We claim that if $s, t > 0$, then

$$\text{(5.27)} \qquad \frac{s}{s+t} R^*_{\{\psi>t+s\}} \xi \le \gamma_t \le R^*_{\{\psi>t\}} \xi \, .$$

Note that first letting $s \to \infty$ and then $t \to \infty$ in (5.27) yields the conclusion of (5.26). For the first inequality in (5.27), because $\gamma_t = R(\xi - tm)$, $\xi \le \gamma_t + tm$. Therefore, using $(s+t)m \le \xi$ on $\{\psi > s+t\}$ for the second inequality,

$$R^*_{\{\psi>t+s\}}\xi \le t \, R^*_{\{\psi>t+s\}} \, m + \gamma_t \le \frac{t}{s+t} R^*_{\{\psi>t+s\}} \, \xi + \gamma_t \, ,$$

establishing the first inequality in (5.27). For the second

$$\xi \le t \, 1_{\{\psi\le t\}} \, m + 1_{\{\psi>t\}} \, \xi \le tm + R^*_{\{\psi>t\}}\xi \, ,$$

and therefore $\gamma_t = R(\xi - tm) \le R^*_{\{\psi>t\}}\xi$. ∎

From (5.17), $\delta_t \le m$ and consequently the integral $\int_0^\infty \delta_t \, dt$ in (5.22) is absolutely continuous with respect to m. Thus if

$\gamma_\infty = 0$, then $\xi << m$. Combining this remark with (5.26) establishes the following:

(5.28) **Corollary.** *In the representation* (5.22), $\gamma_\infty = 0$ *if and only if* $\xi << m$ *and* $\downarrow \lim_{t\to\infty} R^*_{\{\psi>t\}}\, \xi = 0$ *where* ψ *is any Borel measurable version of* $d\xi/dm$.

We now can give a more "constructive" description of the strong maximum and minimum of two excessive measures. Suppose first that $\xi, \eta \in \text{Inv}$. Because $\xi \vee \eta = \xi P_t \vee \eta P_t \leq (\xi \vee \eta)P_t$, we have $\rho := \uparrow \lim_{t\to\infty} (\xi \vee \eta)P_t$ exists and $\rho \leq \xi + \eta$ and hence is σ-finite. Clearly $\rho \in \text{Inv}$. But $\xi \leq (\xi \vee \eta)P_t$ and so $\xi \leq \rho$. However when restricted to Inv the simple and strong orders agree; that is, if $\gamma_1, \gamma_2 \in \text{Inv}$ and $\gamma_1 \leq \gamma_2$, then, because $\gamma_2 - \gamma_1 \in \text{Inv}$, $\gamma_1 \prec \gamma_2$. Consequently $\xi \prec \rho$ and similarly $\eta \prec \rho$. Now suppose $\xi \prec \gamma$ and $\eta \prec \gamma$ with $\gamma \in \text{Exc}$. We want to show $\rho \prec \gamma$ and because γ_i —the invariant part of γ —strongly dominates ξ and η we may suppose $\gamma \in \text{Inv}$. But $\xi \vee \eta \leq \gamma$ and hence $(\xi \vee \eta)P_t \leq \gamma P_t \leq \gamma$. Therefore $\rho \leq \gamma$ and so $\rho \prec \gamma$, ρ and γ being invariant. Thus $\rho = \xi \overset{s}{\vee} \eta$. A similar but simpler argument shows that $(\xi \wedge \eta)_i = \xi \overset{s}{\wedge} \eta$.

Next suppose $\xi, \eta \in \text{Pur}$ and let (μ_t) and (ν_t) be the entrance laws representing ξ and η respectively as in (5.25). Following Neveu [**N61**], note that if $0 < r < s < t$,

$$\begin{aligned}(\mu_s \vee \nu_s)P_{t-s} &= (\mu_r P_{s-r} \vee \nu_r P_{s-r})P_{t-s} \\ &\leq (\mu_r \vee \nu_r)P_{s-r}P_{t-s} = (\mu_r \vee \nu_r)P_{t-r}.\end{aligned}$$

Therefore

$$\lambda_t := \uparrow \lim_{s\downarrow 0} (\mu_s \vee \nu_s)P_{t-s}, \quad t > 0 \tag{5.29}$$

exists and is σ-finite since $\lambda_t \leq \mu_t + \nu_t$. Also

$$\lambda_t P_s = [\uparrow \lim_{r\downarrow 0} (\mu_r \vee \nu_r)P_{t-r}]P_s = \lambda_{t+s},$$

and so $(\lambda_t;\ t > 0)$ is an entrance law. Let $\lambda = \int\limits_0^\infty \lambda_t\,dt$. Since $\lambda \leq \xi + \eta$, $\lambda \in \text{Exc}$ and, of course, $\lambda \in \text{Pur}$. Because $\lambda_t \geq \mu_t$ and $\lambda_t \geq \nu_t$ for each $t > 0$, $(\lambda_t - \mu_t)$ and $(\lambda_t - \nu_t)$ are also entrance laws, and it follows that $\xi \prec \lambda$ and $\eta \prec \lambda$. Suppose $\xi \prec \gamma$ and $\eta \prec \gamma$. We may suppose $\gamma \in \text{Pur}$. If $\xi + \xi_1 = \gamma$ and $\eta + \eta_1 = \gamma$, then $\xi_1, \eta_1 \in \text{Pur}$. Let (μ_t'), (ν_t') and (γ_t) be the entrance laws representing ξ_1, ξ_2, and γ respectively. Since the representation (5.25) is unique, $\mu_t + \mu_t' = \gamma_t$ and $\nu_t + \nu_t' = \gamma_t$. Therefore $\mu_t \vee \nu_t \leq \gamma_t$ and consequently $\lambda_t \leq \gamma_t$. Hence $\lambda \prec \gamma$, that is, $\lambda = \xi \curlyvee \eta$. A similar argument shows that

$$\text{(5.30)} \qquad \theta_t := \downarrow \lim_{s \downarrow 0} (\mu_s \wedge \nu_s) P_{t-s}\,, \quad t > 0$$

defines an entrance law and that $\xi \curlywedge \eta = \int\limits_0^\infty \theta_t\,dt$.

Finally if $\xi, \eta \in \text{Exc}$, then $\xi_i \curlyvee \eta_i + \xi_p \curlyvee \eta_p$ strongly dominates both ξ and η. If $\xi \prec \gamma$ and $\eta \prec \gamma$, then $\xi_i \curlyvee \eta_i \prec \gamma_i$ and $\xi_p \curlyvee \eta_p \prec \gamma_p$. Consequently $\xi \curlyvee \eta = \xi_i \curlyvee \eta_i + \xi_p \curlyvee \eta_p$ and the corresponding statement holds for the strong minima. Thus we have proved the following:

(5.31) **Proposition.** *If* $\xi, \eta \in \text{Inv}$, *then* $\xi \curlyvee \eta = \uparrow \lim\limits_{t \to \infty} (\xi \vee \eta) P_t$ *and* $\xi \curlywedge \eta = (\xi \wedge \eta)_i$. *If* $\xi, \eta \in \text{Pur}$ *with representing entrance laws* (μ_t) *and* (ν_t) *respectively, then* (λ_t) *and* (θ_t) *defined in* (5.29) *and* (5.30) *respectively are entrance laws and*

$$\xi \curlyvee \eta = \int_0^\infty \lambda_t\,dt\,, \quad \xi \curlywedge \eta = \int_0^\infty \theta_t\,dt\,.$$

In the general case $\xi \curlyvee \eta = \xi_i \curlyvee \eta_i + \xi_p \curlyvee \eta_p$ *and* $\xi \curlywedge \eta = \xi_i \curlywedge \eta_i + \xi_p \curlywedge \eta_p$.

Remarks. The same recipe may be used to construct the strong supremum or infimum of an arbitrary upper bounded subset of Exc. It is easy to give an example of two invariant measures ξ and η such that $\xi \wedge \eta \in \text{Pot}$.

In section 2 we gave two Riesz type decompositions of an excessive measure ξ. Namely $\xi = \xi_i + \xi_p$ and $\xi = \xi_c + \xi_d$. We shall now give a third.

(5.32) **Definition.** An excessive measure is *harmonic* provided it strongly dominates no nonzero potential.

Not surprisingly, Har denotes the class of harmonic excessive measures. Thus $\xi \in \text{Har}$ provided $\xi \in \text{Exc}$ and $\mu U \prec \xi$ implies that $\mu U = 0$. The reader should contrast this with the definition of Con in §2. The following result (with a different proof) first appeared in [**GG84**].

(5.33) **Theorem.** *Each* $\xi \in \text{Exc}$ *may be written uniquely as* $\xi = \eta + \mu U$ *with* $\eta \in \text{Har}$.

Proof. Fix $\xi \in \text{Exc}$ and choose $0 < f \leq 1$ with $\xi(f) < \infty$. If ξ is not harmonic, then $\xi = \xi_1 + \mu_1 U$ with $\xi_1 \in \text{Exc}$ and $\mu_1 U \neq 0$, so $\xi_1(f) < \xi(f)$. Suppose that β is a countable ordinal and that for each ordinal $\alpha < \beta$ we have $\xi = \xi_\alpha + \mu_\alpha U$ with $\xi_\alpha \in \text{Exc}$ such that $\xi_\gamma \geq \xi_\alpha$ and $\xi_\gamma(f) > \xi_\alpha(f)$ if $\gamma < \alpha$, and $\mu_\alpha = \sum_{\gamma<\alpha} \mu_\gamma$. If β is not a limit ordinal, let $\alpha = \beta - 1$. If ξ_α is not harmonic we may write $\xi_\alpha = \xi_\beta + \nu U$ with $\xi_\beta(f) < \xi_\alpha(f)$. Then $\xi_\beta \leq \xi_\alpha$ and $\xi = \xi_\beta + (\nu + \mu_\alpha)U$. Defining $\mu_\beta = \nu + \mu_\alpha$, the above assertions now hold for all ordinals $\alpha \leq \beta$. If β is a limit ordinal, define $\mu_\beta = \sum_{\alpha<\beta} \mu_\alpha$ and $\xi_\beta = \inf_{\alpha<\beta} \xi_\alpha$. There exists a sequence (α_n) with $\alpha_n \uparrow \beta$ so that $\xi_{\alpha_n} \downarrow \xi_\beta$. Since $\xi = \xi_{\alpha_n} + \left(\sum_{\gamma<\alpha_n} \mu_\gamma\right) U$, letting $n \to \infty$ we find $\xi = \xi_\beta + \mu_\beta U$. Thus we have constructed ξ_β and $\mu_\beta U$ when β is a limit ordinal. Since $\alpha \to \xi_\alpha(f)$ is strictly decreasing there must exist a countable ordinal α_0 such that $\xi_{\alpha_0} \in \text{Har}$. Hence $\xi = \xi_{\alpha_0} + \mu_{\alpha_0} U$ with $\xi_{\alpha_0} \in \text{Har}$ establishing the existence of the decomposition in (5.33). It remains to prove the uniqueness. Suppose $\xi = \eta_1 + \mu_1 U = \eta_2 + \mu_2 U$ with $\eta_1, \eta_2 \in \text{Har}$. Using (5.13), there exist $\eta_{ij} \in \text{Exc}$ with

$\eta_1 = \eta_{11} + \eta_{12}$, $\eta_2 = \eta_{11} + \eta_{21}$, $\mu_1 U = \eta_{21} + \eta_{22}$, and $\mu_2 U = \eta_{12} + \eta_{22}$, Then (5.23) implies that $\eta_{12}, \eta_{21}, \eta_{22} \in \mathrm{Pot}$ and because η_1 and η_2 are harmonic, this forces $\eta_{12} = \eta_{21} = 0$. Hence $\mu_1 U = \eta_{22} = \mu_2 U$ and $\eta_1 = \eta_{11} = \eta_2$. ∎

(5.34) **Proposition.** Har *is closed under addition. If* $\xi \in \mathrm{Exc}$ *and* $\xi = \eta + \mu U$ *with* $\eta \in \mathrm{Har}$, *then*

$$\begin{aligned}\eta &= \vee\{\gamma \in \mathrm{Har}: \gamma \prec \xi\} \\ &= \curlyvee \{\gamma \in \mathrm{Har}: \gamma \prec \xi\} = \vee\{\gamma \in \mathrm{Har}: \gamma \leq \xi\},\end{aligned}$$

where the first and third suprema are in the natural order and the second is in the strong order.

Proof. Let $\xi_1, \xi_2 \in \mathrm{Har}$ and $\xi = \xi_1 + \xi_2$. Let $\xi = \eta + \mu U$ with $\eta \in \mathrm{Har}$. By (5.13) there exist $\eta_{ij} \in \mathrm{Exc}$ with $\xi_i = \eta_{i1} + \eta_{i2}$, $i = 1, 2$ and $\eta_{12} + \eta_{22} = \mu U$. Hence $\eta_{12}, \eta_{22} \in \mathrm{Pot}$ and since $\xi_i \in \mathrm{Har}$, it follows that $\eta_{12} = \eta_{22} = 0$. Thus $\mu U = 0$ and $\xi = \eta \in \mathrm{Har}$.

Fix $\xi \in \mathrm{Exc}$ and let $H = \{\gamma \in \mathrm{Har}: \gamma \prec \xi\}$. If $\xi = \eta + \mu U$ is the decomposition of ξ into its harmonic and potential parts, then $\eta \in H$. If $\gamma \in H$ then $\xi = \gamma + \lambda$ with $\lambda \in \mathrm{Exc}$. Writing $\lambda = \rho + \pi$ with $\rho \in \mathrm{Har}$ and $\pi \in \mathrm{Pot}$ the uniqueness of the decomposition (5.33) and the closure of Har under addition imply $\eta = \gamma + \rho$. Therefore $\gamma \prec \eta$. It now follows that $\eta = \curlyvee H$, and since $\eta \in H$ and $\vee H \leq \curlyvee H$, the first two expressions for η are established. For the third suppose $\gamma \in \mathrm{Har}$ and $\gamma \leq \xi = \eta + \mu U$. By (5.9), $\gamma = \gamma_1 + \gamma_2$ with $\gamma_1 \leq \eta$ and $\gamma_2 \leq \mu U$. Then $\gamma_2 \in \mathrm{Pot}$ by (5.23) and hence $\gamma_2 = 0$. Thus $\gamma = \gamma_1 \leq \eta$ and so $\eta = \vee\{\gamma \in \mathrm{Har}: \gamma \leq \xi\}$. ∎

(5.35) **Corollary.** *Let* $\xi \in \mathrm{Exc}$. *Then* $\xi \in \mathrm{Pot}$ *if and only if the only harmonic measure that* ξ *dominates in either order is zero. If* $\xi = \eta + \mu U$ *with* $\eta \in \mathrm{Har}$, *then* μU *is the largest element of* Pot *(in either order) that is strongly dominated by* ξ.

Proof. Suppose $\xi \in \mathrm{Pot}$, $\eta \in \mathrm{Har}$, and $\eta \leq \xi$. By (5.23), $\eta \in \mathrm{Pot}$ and hence $\eta = 0$. Suppose $\xi \in \mathrm{Exc}$ and ξ dominates no nonzero harmonic measure in the natural order. Then according to (5.34) the harmonic part of $\xi = 0$, so $\xi \in \mathrm{Pot}$. If $\xi = \eta + \mu U$ with $\eta \in \mathrm{Har}$ and $\pi \in \mathrm{Pot}$ with $\pi \prec \xi$, then using (5.13) it is easy to check that $\pi \prec \mu U$. ■

A subcone C of Exc is *solid* in the natural order (resp. strong order) provided that $\xi \in \mathrm{Exc}$, $\eta \in C$, and $\xi \leq \eta$ (resp. $\xi \prec \eta$) imply $\xi \in C$. It is immediate from the properties developed in §2 and this section, that Con, Dis, Pur and Pot are solid in the natural order and that Inv and Har are solid in the strong order. One also has the following inclusions:

$$\mathrm{Con} \subset \mathrm{Inv} \subset \mathrm{Har}\,; \quad \mathrm{Pot} \subset \mathrm{Pur} \subset \mathrm{Dis}\,.$$

We end this section with a famous example of Helms and Johnson of an excessive measure that is in $\mathrm{Har} \cap \mathrm{Pur}$. Let X be Brownian motion in $\mathbb{R}^3$ and let Y be the restriction of X to the absorbing set $\mathbb{R}^3 \backslash \{0\}$. Let $\lambda(dx) = |x|^{-1}\, dx$ where dx is Lebesgue measure in $\mathbb{R}^3$. Then $\lambda \in \mathrm{Pur}\,(Y)$ being the integral of the entrance law (for Y), $cP_t(0, dx)$ where $c > 0$ is an appropriate constant and (P_t) is the Brownian semigroup on $\mathbb{R}^3$. If $\pi \in \mathrm{Pot}\,(Y)$ and $\pi \prec \lambda$, then π and λ may be regarded as excessive measures of X, and as such $\pi = c'\epsilon_0 U$ for some $c' \geq 0$ where U is the potential kernel for X. But $\pi \in \mathrm{Pot}\,(Y)$ implies $\pi = \mu U$ for a measure μ carried by $\mathbb{R}^3 \backslash \{0\}$. Therefore (2.12) implies that $c' = 0$. Hence $\lambda \in \mathrm{Har}(Y)$.

6. Kuznetsov Measures

In this section we are going to introduce a very powerful tool for the study of excessive measures. Unfortunately it is necessary to make a supplementary hypothesis (see (6.2) below) on our right process X. However, before coming to that we introduce a generalization of an entrance law as defined in (5.24).

(6.1) **Definition.** An *entrance rule* for (P_t) (or X) is a family $(\eta_t)_{t\in\mathbb{R}}$ of σ-finite measures on E such that $\eta_s P_{t-s} \leq \eta_t$ for $s < t$ and $\eta_s P_{t-s} \uparrow \eta_t$ as $s \uparrow t$.

An entrance rule as defined in (6.1) is called a regular entrance rule in [**DM**, XVIII] but this is the only type we shall need. We emphasize that η_t is indexed by all real t not just $t \geq 0$. If $\xi \in \text{Exc}$ then defining $\eta_t = \xi$ for all $t \in \mathbb{R}$ it is obvious that (η_t) is an entrance rule. If $\nu = (\nu_t,\ t > 0)$ is an entrance law as defined in (5.24), then extending ν by setting $\nu_t = 0$ if $t \leq 0$, it is easily checked that the extended ν is an entrance rule. More generally an *entrance law at* u, $-\infty \leq u < \infty$ is an entrance rule (η_t) such that $\eta_t = 0$ if $t \leq u$ and $\eta_t P_s = \eta_{t+s}$ if $t > u$ and $s > 0$. Clearly an entrance law as defined in (5.24) and extended as above is an entrance law at 0. Thus entrance laws are regarded as a subclass of entrance rules. Note that if $\xi \in \text{Inv}$, then $\eta_t = \xi$ for all $t \in \mathbb{R}$ defines an entrance law at $-\infty$. If (η_t) is an entrance rule and $r < s < t$, then $(\eta_r P_{s-r})P_{t-s} \leq \eta_s P_{t-s}$ so $s \to \eta_s P_{t-s}$ is increasing on $]-\infty, t[$.

We now introduce an additional axiom on X. This is Axiom 3 in [**DM**, XVI-32].

(6.2) (i) *For each* $q \geq 0$ *the* q*-excessive functions are nearly Borel.*

(ii) *For each probability* μ *on* E *there exists a subset* $E_\mu \subset E$ *which is Lusinian and such that* $P^\mu[X_t \notin E_\mu$ *for some* $t \geq 0] = 0$.

Both statements in (6.2) refer to the original topology on E. Recall that E_μ being Lusinian means that in its topology as a subspace of E, it is homeomorphic to a Borel subspace of a compact metric space. Of course, the q-excessive functions are always nearly Borel in the Ray topology, but even if E is Lusinian in the original topology it need not be so in the Ray topology. If X is a Borel right process; that is, E is Lusinian and (P_t) maps Borel functions into Borel functions, then it is well-known that (6.2) holds with $E_\mu = E$ for all μ in (6.2ii). See, for example, [**G**, (9.4)]. The importance of (6.2i) is that it guarantees that every right continuous (in the original topology of E) realization of (P_t) is a right process. See [**S**, (19.3)]. In the remainder of this section it is assumed that (6.2) is in force. Assuming (6.2) is more flexible than assuming that X is a Borel right process because it is stable under most familiar transformations of X. The main problem is that it is sometimes difficult to verify.

We now introduce the notation necessary to describe Kuznetsov measures. Let W be the space of paths $w: \mathbb{R} \to E \cup \{\Delta\}$ that are E-valued and right continuous on an open interval $]\alpha(w), \beta(w)[$ and take the value Δ elsewhere. The path $[\Delta]$ constantly equal to Δ corresponds to $]\alpha, \beta[$ being empty and we set $\alpha([\Delta]) = +\infty$, $\beta([\Delta]) = -\infty$. Let $Y = (Y_t;\ t \in \mathbb{R})$ denote the coordinate process on W, $Y_t(w) = w(t)$ and set $Y_\infty(w) = Y_{-\infty}(w) = \Delta$. Put $\mathcal{G}^0 = \sigma\{Y_t;\ t \in \mathbb{R}\}$ and $\mathcal{G}^0_t = \sigma\{Y_s;\ s \leq t\}$. The shifts $\sigma_t: W \to W$ are defined by $Y_s \circ \sigma_t = Y_{t+s}$. Clearly $\sigma_t \in \mathcal{G}^0_{s+t}|\mathcal{G}^0_s$ for all $s, t \in \mathbb{R}$. The following theorem is proved in [**DM**, XVIII]. See also [**GG87**], [**Ku74**], and [**Ku84**].

(6.3) **Theorem.** *Assume* (6.2) *and let* $\nu = (\nu_t)$ *be an entrance rule. Then there exists a unique measure* Q_ν *on* $(W, \mathcal{G}^0)$ *not charging* $\{[\Delta]\}$ *such that if* $t_1 < \cdots < t_n$,

$$(6.4)\quad Q_\nu(\alpha < t_1, Y_{t_1} \in dx_1, \ldots, Y_{t_n} \in dx_n, t_n < \beta)$$
$$= \nu_{t_1}(dx_1) P_{t_2 - t_1}(x_1, dx_2) \cdots P_{t_n - t_{n-1}}(x_{n-1}, dx_n).$$

Moreover Q_ν *is* σ*-finite.*

If $m \in \text{Exc}$ and we put $\nu_t = m$ for all t, we write Q_m in place of Q_ν. In this case because only differences of t values appear on the right hand side of (6.4), it is evident that Q_m is stationary; that is, $\sigma_t(Q_m) = Q_m$ for each $t \in \mathbb{R}$. We call Q_ν (or Q_m) the *Kuznetsov* measure of ν (or m). It is immediate from (6.4) that for each $s \in \mathbb{R}$ the process $(Y_{s+t};\ t \geq 0)$ under Q_ν restricted to $\{Y_s \in E\} = \{\alpha < s < \beta\}$ has the same law as $(X_t;\ t \geq 0)$ under P^{ν_s}. Consequently one may suppose, if one wishes, that on $]\alpha(w), \beta(w)[$, $t \to w(t)$ is right continuous in the Ray topology and has left limits in a Ray compactification $\overline{E}$ of E. See for example, [**S**, §18] or [**G**, (11.2)]. However, this is not always convenient, especially when dealing with several processes simultaneously. We shall state explicitly when, if ever, we make this additional assumption on W.

It is shown in [**DM**, XVIII] that (Y_t) is strong Markov under Q_ν in the following sense. Let $\mathcal{G}^\nu$ be the Q_ν completion of $\mathcal{G}^0$ and let $\mathcal{G}_t^\nu$ be the σ-algebra generated by $\mathcal{G}_t^0$ and the ideal of Q_ν null sets in $\mathcal{G}^\nu$. Of course, we write $\mathcal{G}^m$, etc. when $m \in \text{Exc}$ and $\nu_t = m$ for all t. The filtration $(\mathcal{G}_t^m)$ is right continuous when $m \in \text{Exc}$. A $(\mathcal{G}_{t+}^\nu)$ stopping time T is a map $T\colon W \to [-\infty, \infty]$ such that $\{T < t\} \in \mathcal{G}_t^\nu$ for all $t \in \mathbb{R}$, and $\mathcal{G}_{T+}^\nu$ it the σ-algebra consisting of all sets $\Lambda \in \mathcal{G}^\nu$ such that $\Lambda \cap \{T < t\} \in \mathcal{G}_t^\nu$ for all $t \in \mathbb{R}$. Of course, stopping times relative to $(\mathcal{G}_t^m)$, $(\mathcal{G}_{t+}^0)$, etc. are defined similarly. Of special importance are the σ-algebras $\mathcal{G}^*$ and $\mathcal{G}_t^*$, the universal completions of $\mathcal{G}^0$ and $(\mathcal{G}_t^0)$ respectively. In addition, the σ-algebras $\mathcal{G} := \bigcap_m \mathcal{G}^m$ and $\mathcal{G}_t := \bigcap_m \mathcal{G}_t^m$ where the intersection is over all $m \in \mathrm{Exc}$ are useful. They are analogous to $\mathcal{F}$ and $\mathcal{F}_t$ for the process X. Clearly the filtration $(\mathcal{G}_t)$ is right continuous.

Here is one form of the strong Markov property; another form is given in (6.6). See also (6.15). Let ν be an entrance rule and T a $(\mathcal{G}^\nu_{t+})$ stopping time. Then the process $(Y_{T+t};\, t \ge 0)$ on $\{\alpha < T < \beta\}$ under the measure Q_ν is a strong Markov process in the usual sense relative to the filtration $(\mathcal{G}^\nu_{T+t};\, t \ge 0)$ and having (P_t) as its semigroup. Moreover Q_ν is σ-finite on the trace of $\mathcal{G}^\nu_{T+}$ on $\{\alpha < T < \beta\}$. See [**DM**, XVIII].

We now describe a convenient realization of (P_t) as a right process that we shall use in the sequel. First we define birthing, killing, and truncated shift operators on W:

$$
\begin{aligned}
&b_t w(s) = w(s) \quad \text{if} \quad s > t,\ b_t w(s) = \Delta \quad \text{if} \quad s \le t\,;\\
(6.5)\quad &k_t w(s) = w(s) \quad \text{if} \quad s < t,\ k_t w(s) = \Delta \quad \text{if} \quad s \ge t\,;\\
&\theta_t w(s) = w(t+s) \quad \text{if} \quad s > 0,\ \theta_t w(s) = \Delta \quad \text{if} \quad s \le 0\,.
\end{aligned}
$$

Note that $\theta_t = b_0 \sigma_t = \sigma_t b_t$ and hence $\theta_t \circ \sigma_s = \theta_{t+s}$. Define

$$\Omega = \{w \in W\colon \alpha(w) = 0,\ Y_{\alpha+}(w) \quad \text{exists in} \quad E\} \cup \{[\Delta]\}\,.$$

Observe that $\theta_t(\{\alpha < t\}) \subset \Omega$. Let X_t be the restriction of Y_t to Ω if $t > 0$ and $X_0(\omega) = Y_{0+}(\omega)$ for $\omega \in \Omega$. Then $t \to X_t(\omega)$ is right continuous for $t \ge 0$ and $\zeta(\omega) := \inf\{t \ge 0\colon X_t(\omega) = \Delta\} = \beta(\omega)$ provided $\omega \ne [\Delta]$. Clearly $\zeta \circ \theta_t = \beta - t$ if $\alpha < t < \beta$. If $\mathcal{F}^0 := \sigma\{X_t;\ t \ge 0\}$ and $\mathcal{F}^0_t = \sigma\{X_s\colon s \le t\}$, then $\mathcal{F}^0 = \mathcal{G}^0|_\Omega$, $\mathcal{F}^0_t = \mathcal{G}^0_t|_\Omega$ if $t > 0$, and $\mathcal{F}^0_0 = \sigma(X_0) \subset \mathcal{G}^0_{0+}|_\Omega$. It is evident that $X_s \circ \theta_t = X_{s+t}$ on Ω for $s, t \ge 0$ and that $X_s \circ \theta_t = Y_{s+t}$ if $s \ge 0$, $\alpha < t < \infty$. Consequently for each $s \ge 0$, $\theta_t \in (\mathcal{G}^0_{t+s}|_{\{\alpha<t\}})|\mathcal{F}^0_s$. Clearly we may identify Ω with the canonical space of right continuous paths from $[0,\infty[$ to $E \cup \{\Delta\}$ having Δ as cemetery point and $(X_t;\, t \ge 0)$ as the coordinate process on Ω so identified. Because of (6.2) there exist probabilities P^x, $x \in E$ on $(\Omega, \mathcal{F}^0)$ so that $X = (\Omega, \mathcal{F}, \mathcal{F}_t, X_t, \theta_t, P^x)$ is a realization of (P_t) as a

right process. See [**S**, (19.3)]. Of course, $\mathcal{F}$, $\mathcal{F}_t$, etc. have their usual meanings relative to this realization X of (P_t). From now on whenever (6.2) is in force, X refers to this particular realization. The strong Markov property for Q_ν now takes the following form. Let $\nu = (\nu_t)$ be an entrance rule, T a $(\mathcal{G}^\nu_{t+})$ stopping time, $G \in p\mathcal{G}^\nu_{T+}$ and $F \in p\mathcal{F}^*$ —recall $\mathcal{F}^*$ is the universal completion of $\mathcal{F}^0$ —then

$$\text{(6.6)} \qquad Q_\nu(F \circ \theta_T G; \alpha < T < \beta) \\ = Q_\nu(G P^{Y(T)}(F); \alpha < T < \beta),$$

and, as mentioned before, Q_ν is σ-finite on the trace of $\mathcal{G}^\nu_{T+}$ on $\{\alpha < T < \beta\}$.

Suppose $\nu = (\nu_t)$ is an entrance law at 0, then $Q_\nu(\alpha \neq 0) = 0$. To see this note that $\{\alpha < 0\} \subset \cup\{\alpha < r < \beta\}$ where the union is over all rationals $r < 0$. But $Q_\nu(\alpha < r < \beta) = \nu_r(1) = 0$ if $r < 0$. Because Q_ν doesn't charge $\{[\Delta]\}$ one has a.s. Q_ν, $\{0 < \alpha\} = \bigcup_{0<r<s} \{r \le \alpha < s < \beta\}$ where again r and s are rational. If $f > 0$ with $\nu_s(f) < \infty$, then for $0 < r < s$

$$Q_\nu(r \le \alpha < s < \beta, f \circ Y_s) = Q_\nu(f \circ Y_s) - Q_\nu(\alpha < r, f \circ Y_s) \\ = \nu_s(f) - \nu_r P_{s-r}(f) = 0,$$

where, of course, we use the usual convention $f(\Delta) = 0$. Thus $Q_\nu(\alpha \neq 0) = 0$. Similarly if ν is an entrance law at u, then $Q_\nu(\alpha \neq u) = 0$. In particular if $\xi \in \text{Inv}$, $Q_\xi(\alpha > -\infty) = 0$. Dually it is easily checked that if $\xi \in \text{Exc}$, then $Q_\xi(\beta < \infty) = 0$ if and only if $P_t 1 = 1$ a.e. ξ for each t. In particular, because of (2.4) and (2.5), $\alpha = -\infty$ and $\beta = +\infty$ a.s. Q_ξ whenever $\xi \in \text{Con}$.

Let $m \in \text{Exc}$ and suppose $S: W \to [-\infty, \infty]$ is $\mathcal{G}^m$ measurable and satisfies $S \circ \sigma_t = S - t$ a.s. Q_m for each $t \in \mathbb{R}$. It is then easy to check that $\gamma(dt) := Q_m(S \in dt)$ defines a translation invariant measure on $\mathbb{R}$ which is a countable sum of finite measures. As such it is a multiple of Lebesgue measure, $\gamma(dt) = c\,dt$ where $0 \le c \le \infty$. See the proof of (8.23) immediately following (8.25), for example. In particular $Q_m(S = t) = 0$ for each fixed t. Of course α and β satisfy the hypotheses on S.

The next result is due to Fitzsimmons and Maisonneuve [**FM86**].

(6.7) **Theorem.** *Let* $m \in \text{Exc}$. *Then* $m \in \text{Inv}$ *(resp.* Pur*) if and only if* $Q_m(\alpha > -\infty) = 0$ *(resp.* $Q_m(\alpha = -\infty) = 0$*). If* $m = m_i + m_p$ *is the decomposition of* m *into its invariant and purely excessive parts, then* $m_i(f) = Q_m[f \circ Y_0;\ \alpha = -\infty]$ *and* $m_p(f) = Q_m[f \circ Y_0;\ \alpha > -\infty]$. *Define for* $t > 0$

$$\nu_t(f) = Q_m[f \circ Y_{\alpha+t};\quad 0 < \alpha < 1]. \tag{6.8}$$

Then (ν_t) *is an entrance law and* $m_p = \int_0^\infty \nu_t\, dt$.

Proof. Let $f \in p\mathcal{E}$ with $m(f) < \infty$. Recall that $f(\Delta) = 0$ by convention. Then for $t > 0$

$$\begin{aligned} m(P_t f) &= Q_m[P_t f \circ Y_0] = Q_m[f \circ Y_t;\ \alpha < 0] \\ &= Q_m[f \circ Y_0;\ \alpha < -t] \to Q_m[f \circ Y_0;\ \alpha = -\infty] \end{aligned}$$

as $t \to \infty$, where we have used the Markov property for the second equality and the stationarity of Q_m for the third. The assertions in the second and third sentences of (6.7) now are immediate. With f as above and $s, t > 0$

$$\begin{aligned} \nu_t(P_s f) &= Q_m[P_s f \circ Y_{\alpha+t};\ 0 < \alpha < 1] \\ &= Q_m[f \circ Y_{\alpha+t+s};\ 0 < \alpha < 1] = \nu_{t+s}(f), \end{aligned} \tag{6.9}$$

where the second equality comes from the strong Markov property applied to the $(\mathcal{G}^0_{t+})$ stopping time $\alpha + t$ and the fact that

$X_s \circ \theta_{\alpha+t} = Y_{\alpha+s+t}$ if $s > 0$, $t > 0$ and $\alpha \in \mathbb{R}$. It now follows that $t \to \nu_t(f)$ is Borel measurable. Therefore

$$(6.10)\int_0^\infty \nu_t(f)\,dt = Q_m\left[\int_0^\infty f\circ Y_{\alpha+t}\,dt;\ 0<\alpha<1\right]$$

$$= Q_m\left[\int_{-\infty}^\infty f\circ Y_t\, 1_{]\alpha,\infty[}(t)\,dt;\ 0<\alpha<1\right]$$

$$= \int_{-\infty}^\infty Q_m[f\circ Y_t;\ 0<\alpha<1]\,dt$$

$$= \int_{-\infty}^\infty Q_m[f\circ Y_0;\ -t<\alpha<1-t]\,dt$$

$$= Q_m[f\circ Y_0;\ \alpha>-\infty] = m_p(f)\,.$$

Moreover, using (6.9), $\nu_t U(f) = \int_t^\infty \nu_s(f)\,ds \le m_p(f) < \infty$, and so ν_t is σ-finite. Consequently from (6.9), (ν_t) is an entrance law and $m_p = \int_0^\infty \nu_t\,dt$ according to (6.10). ■

Remark. Of course, because of the stationarity of Q_m one has $m_i(f) = Q_m(f\circ Y_t;\ \alpha = -\infty)$ and $m_p(f) = Q_m(f\circ Y_t;\ \alpha > -\infty)$ for any $t \in \mathbb{R}$.

Next we are going to describe Har and Pot in terms of the associated Kuznetsov measure, still following [**FM86**]. We shall need the following lemma.

(6.11) **Lemma**. *Suppose* $m \in \mathrm{Pur}$ *and* $\nu = (\nu_t)$ *is the entrance law with* $m = \int_0^\infty \nu_t\,dt$. *Then* $Q_m = \int_{-\infty}^\infty \sigma_t(Q_\nu)\,dt$.

Proof. Let $t_1 < t_2 < \cdots < t_n$. Then

$$\sigma_t(Q_\nu)(\alpha < t_1, Y_{t_1} \in dx_1, \ldots, Y_{t_n} \in dx_n,\ t_n < \beta)$$
$$= \nu_{t+t_1}(dx_1)P_{t_2-t_1}(x_1,dx_2)\cdots P_{t_n-t_{n-1}}(x_{n-1},dx_n)\,.$$

Integrating this relationship in t over $\mathbb{R}$ and using the fact that $\nu_t = 0$ if $t \leq 0$, the conclusion of (6.11) follows from the uniqueness assertion in (6.3). ∎

In order to proceed we need to introduce certain subsets of W on which Y_t has a good behavior at α. To this end we fix a countable, uniformly dense subset, D, of the bounded real valued uniformly continuous (relative to some metric on E compatible with the topology of E and such that the completion of E in this metric is compact) functions on E. Given $h \in \mathcal{E}$ with $0 < h \leq 1$ define $W(h)$ by the following conditions: (i) $\alpha \in \mathbb{R}$, (ii) $Y_{\alpha+}$ exists in E, (iii) $U^r g \circ Y_{\alpha+1/n} \to U^r g \circ Y_{\alpha+}$ as $n \to \infty$ for each $g \in D$ and each strictly positive rational r, (iv) $Uh \circ Y_{\alpha+1/n} \to Uh \circ Y_{\alpha+}$ as $n \to \infty$. Note that $W(h)$ depends on h only through condition (iv). Recall that $\mathcal{G}^*$ (resp. $\mathcal{G}_t^*$) is the universal completion of $\mathcal{G}^0$ (resp. $\mathcal{G}_t^0$). Then $\mathcal{G}^* \subset \mathcal{G}$ and $\mathcal{G}_{t+}^* \subset \mathcal{G}_t$ for all t. Standard considerations using the fact that E is a Radon space (see, for example, [**DM**, IV-T33]) show that $W(h) \in \mathcal{G}_{\alpha+}^*$. Of course, α is a stopping time relative to $(\mathcal{G}_{t+}^0)$ and, hence, relative to $(\mathcal{G}_{t+}^*)$. Clearly both $W(h)$ and $W(h)^c$ are invariant; $\sigma_t^{-1} W(h) = W(h)$, $\sigma_t^{-1} W(h)^c = W(h)^c$ for $t \in \mathbb{R}$. Define

$$\text{(6.12)}\quad \begin{aligned} Y_t^*(w) &= Y_t(w) \quad \text{if} \quad t \neq \alpha(w) \quad \text{or} \quad w \notin W(h) \\ &= Y_{\alpha+}(w) \quad \text{if} \quad t = \alpha(w) \quad \text{and} \quad w \in W(h). \end{aligned}$$

We are going to show that if $m \in \text{Exc}$ and $m(h) < \infty$, then Y^* has the strong Markov property at α under Q_m on $W(h)$. To formulate this more precisely define

$$\text{(6.13)}\qquad W_t(h) = \{\alpha < t\} \cup (\{\alpha = t\} \cap W(h)), \quad t \in \mathbb{R}.$$

Of course, if $T \colon W \to [-\infty, \infty]$,

$$\text{(6.14)}\qquad W_T(h) = \{\alpha < T\} \cup (\{\alpha = T\} \cap W(h)).$$

Note that $\theta_t(W_t(h)) \subset \Omega$ and that $X_s \circ \theta_t = Y^*_{s+t}$ for $s \geq 0$ on $W_t(h)$. Here is the extended strong Markov property which is again taken from [**FM86**]. See also [**GSt86**, (5.8)].

(6.15) **Proposition.** *Let* $m \in \text{Exc}$ *and* $m(h) < \infty$. *Let* T *be a* $(\mathcal{G}^m_t)$ *stopping time,* $F \in p\mathcal{F}^*$, *and* $G \in p\mathcal{G}^m_T$. *Then*

$$Q_m[GF \circ \theta_T;\ W_T(h);\ -\infty < \alpha \leq T < \beta]$$
$$= Q_m[GP^{Y^*(T)}(F);\ W_T(h);\ -\infty < \alpha \leq T < \beta].$$

Proof. Suppose first that $m \in \text{Pur}$ and that $m = \int_0^\infty \nu_t\, dt$ where ν is an entrance law. We shall first prove (6.15) with Q_m replaced by Q_ν. That is, since $\alpha = 0$ a.s. Q_ν, we shall prove for T a $(\mathcal{G}^\nu_{t+})$ stopping time, $F \in p\mathcal{F}^*$, and $G \in p\mathcal{G}^\nu_{T+}$ that

$$\text{(6.16)}\quad Q_\nu[GF \circ \theta_T;\ W_T(h);\ 0 \leq T < \beta]$$
$$= Q_\nu[GP^{Y^*(T)}(F);\ W_T(h);\ 0 \leq T < \beta].$$

Using standard techniques it suffices to prove (6.16) for T a $(\mathcal{G}^0_{t+})$ stopping time and $G = 1$. Moreover in light of (6.6) and the fact that $W_T(h) \cap \{\alpha < T\} = \{\alpha < T\}$, it suffices to prove (6.16) with $\{0 \leq T < \beta\}$ replaced by $\{T = 0\}$ on both sides. Also $W_0(h) = W(h)$ a.s. Q_ν. Thus the proof of (6.16) is reduced to showing

$$\text{(6.16*)}\quad Q_\nu[F \circ \theta_0;\ W(h);\ T = 0]$$
$$= Q_\nu[P^{Y^*(0)}(F);\ W(h);\ T = 0].$$

To this end for $f \in pb\mathcal{E}^e$ and $\Lambda \in \mathcal{G}^\nu_{0+}$ define a measure Q^f on $\mathcal{G}^\nu$ by

$$Q^f(H) = Q_\nu[(f \cdot Uh) \circ Y_{0+}H;\ \Lambda \cap W(h)]$$

for $H \in p\mathcal{G}^\nu$. If $f \leq 1$,

$$Q^f(1) \leq Q_\nu[Uh \circ Y_{0+};\ W(h)] \leq \liminf_k Q_\nu[Uh \circ Y_{1/k};\ W(h)]$$

$$\leq \liminf_k \nu_{1/k}(Uh) = \int_0^\infty \nu_t(h)\,dt = m(h) < \infty\,,$$

where the second inequality follows from Fatou's lemma and condition (iv) in the definition of $W(h)$. In particular $Uh \circ Y_{0+} < \infty$ a.s. Q_ν on $W(h)$. In fact, this is the only place that (iv) is ever used. Thus if $f \in bp\mathcal{E}^e$, $g \in D$, and r a positive rational we obtain ($\alpha = 0$ a.s. Q^f)

$$\int_0^\infty e^{-rt} Q^f(g \circ Y_t)\,dt = \lim_{k\to\infty} \int_{1/k}^\infty e^{-rt} Q^f(g \circ X_{t-1/k} \circ \theta_{1/k})\,dt$$

$$= \lim_{k\to\infty} \int_{1/k}^\infty e^{-rt} Q^f(P_{t-1/k}\, g \circ Y_{1/k})\,dt$$

$$= \lim_{k\to\infty} e^{-r/k} Q^f(U^r g \circ Y_{1/k}) = Q^f(U^r g \circ Y_{0+})$$

$$= \int_0^\infty e^{-rt} Q^f(P_t g \circ Y_{0+})\,dt\,,$$

where the second equality uses the Markov property of (Y_t) under Q_ν — $W(h) \in \mathcal{G}^\nu_{\alpha+} = \mathcal{G}^\nu_{0+}$ a.s. Q_ν. By continuity this then holds for all $r > 0$, and since the extremes are the Laplace transforms of bounded right continuous functions of t on $]0,\infty[$, we obtain $Q^f(g \circ Y_t) = Q^f(P_t g \circ Y_{0+})$ for $t > 0$. Recall that on $W(h) \cap \{\alpha = 0\}$, $Y_t^* = Y_t$ if $t > 0$ and $Y_0^* = Y_{0+}$. Writing $Z = (f \cdot Uh)(Y_0^*)$ for notational convenience we obtain by continuity, putting back the definition of Q^f in terms of Q_ν

$$(6.17)\quad Q_\nu[Zg \circ Y_t^*;\ \Lambda \cap W(h)] = Q_\nu[ZP_t g \circ Y_0^*;\ \Lambda \cap W(h)]$$

for all $g \in D$, $t \geq 0$, and $\Lambda \in \mathcal{G}^\nu_{0+}$. But D is dense in $C_u(E)$ —the bounded uniformly continuous functions on E — and hence (6.17) holds for all $g \in C_u(E)$. Finally a monotone class argument extends (6.17) to $g \in b\mathcal{E}$ and then to $g \in b\mathcal{E}^*$. Since $\Lambda \in \mathcal{G}_{0+}^{\nu}$ is arbitrary we may replace 1_Λ by any $\Gamma \in b\mathcal{G}_{0+}^{\nu}$. Recall that both sides of (6.17) are bounded by $m(h)$ if f and g are bounded by 1. Also the right side of (6.17) may be written

$$Q_\nu[Z\Gamma P^{Y^*(0)}(g \circ X_t);\ W(h)],$$

where we have replaced 1_Λ by $\Gamma \in b\mathcal{G}_{0+}^{\nu}$.

Now let $F = g_0 \circ X_0 \prod_{j=1}^{n} g_j \circ X_{t_j}$ where $0 < t_1 < \cdots < t_n$ and $g_j \in b\mathcal{E}^*$, $0 \le j \le n$. Let $\varphi(x) = P^x\left(\prod_{j=1}^{n} g_j \circ X_{t_j - t_1}\right)$. Then $F \circ \theta_0 = g_0 \circ Y_0^* \prod_{j=1}^{n} g_j \circ Y_{t_j}$ on $\{\alpha = 0\} \cap W(h)$, and so

$$\begin{aligned}(6.18)\quad Q_\nu[Z\Gamma F \circ \theta_0;\ W(h)] &= Q_\nu[Z\Gamma g_0 \circ Y_0^* \varphi \circ Y_{t_1};\ W(h)] \\ &= Q_\nu[Z\Gamma g_0 \circ Y_0^* P^{Y^*(0)}(\varphi \circ X_{t_1});\ W(h)] \\ &= Q_\nu[Z\Gamma P^{Y^*(0)}(F);\ W(h)]\end{aligned}$$

where the first equality follows from the Markov property of (Y_t) under Q_ν and the second from (6.17). As usual (6.18) extends to $F \in b\mathcal{F}^*$. Now write (6.18) for $F \ge 0$ and $G \ge 0$ and recall that $Z = f \circ Y_0^* Uh \circ Y_0^*$ with $f \in bp\mathcal{E}^e$. By monotone convergence (6.18) then holds for $f \in p\mathcal{E}^e$, and since $Uh > 0$ we may take $f = (Uh)^{-1}$. (Take $f = 0$ if $Uh = \infty$ and recall that $Uh \circ Y_0^* = Uh \circ Y_{0+} < \infty$ a.s. Q_ν on $W(h)$.) Then (6.16*) follows because $\{T = 0\} \in \mathcal{G}_{0+}^0$. Hence we have established (6.16).

Since $m = \int_0^\infty \nu_t\, dt$, according to (6.11),

$$Q_m = \int_{-\infty}^{\infty} \sigma_t(Q_\nu)\, dt\,.$$

It suffices to prove (6.15) for T a $(\mathcal{G}_{t+}^0)$ stopping time and $G = 1$. It is easily checked that if T is a $(\mathcal{G}_{t+}^0)$ stopping time then so is $T(t) := t + T \circ \sigma_t$ for every $t \in \mathbb{R}$. Moreover $\theta_T \circ \sigma_t = \theta_{T(t)}$ and

$$\sigma_t^{-1}[W_T(h) \cap \{\alpha \leq T < \beta\}] = W_{T(t)}(h) \cap \{\alpha \leq T(t) < \beta\}.$$

Combining these remarks with (6.7) and (6.16) yields (6.15) when $m \in \text{Pur}$. If $m \in \text{Exc}$, then the formula in (6.15) is unchanged if we replace m by m_p and so reduces to the case $m \in \text{Pur}$. ■

We now are prepared to describe Har and Pot in terms of Kuznetsov measures following **[FM86]**.

(6.19) **Theorem.** *Let* $m \in \text{Exc}$ *and* $0 < h \leq 1$ *with* $m(h) < \infty$. *Then* $m \in \text{Har}$ *(resp.* Pot*) if and only if* $Q_m(W(h)) = 0$ *(resp.* $Q_m(W(h)^c) = 0$*). If* $m = \eta + \pi$ *with* $\eta \in \text{Har}$, $\pi \in \text{Pot}$, *then* $Q_\eta = Q_m(\,\cdot\,; W(h)^c)$, $Q_\pi = Q_m(\,\cdot\,; W(h))$ *and* $\pi = \mu U$ *where*

$$\mu(f) = Q_m[f \circ Y_{\alpha+};\ 0 < \alpha < 1;\ W(h)]. \tag{6.20}$$

If $\nu_t = \mu P_t$ *for* $t > 0$, *then* $\nu = (\nu_t; t > 0)$ *is an entrance law and* $(Y_{t+};\ t \geq 0)$ *under* Q_ν *has the same law as* $(X_t;\ t \geq 0)$ *under* P^μ.

Proof. Suppose first $m \in \text{Pot}$, say $m = \mu U$. Let $\nu_t = \mu P_t$ so that $(\nu_t;\ t > 0)$ is an entrance law. It is immediate that $(Y_t;\ t > 0)$ under Q_ν has the same law as $(X_t;\ t > 0)$ under P^μ. Let f be q-excessive, $q \geq 0$. Then f is nearly Borel and P^μ almost surely $t \to f \circ X_t$ and $t \to X_t$ are right continuous on $[0, \infty[$. It follows that almost surely Q_ν, Y_{0+} exists in E and $f \circ Y_t \to f \circ Y_{0+}$ as $t \downarrow 0$. Hence Q_ν is carried by $W(h)$. Therefore

$$Q_m[W(h)^c] = \int_{-\infty}^{\infty} \sigma_t Q_\nu[W(h)^c]\, dt = 0$$

because $W(h)^c$ is invariant. It now is clear that $(Y_{t+},\ t \geq 0)$ under Q_ν has the same law as $(X_t;\ t \geq 0)$ under P^μ.

Next suppose $Q_m[W(h)] = 0$. Let $m = \eta + \mu U$ be the decomposition of m into its harmonic and potential parts. Then $Q_{\mu U} \le Q_m$ and so $Q_{\mu U}[W(h)] = 0$. But $Q_{\mu U}$ is carried by $W(h)$. Therefore $\mu U = 0$ and $m = \eta \in \text{Har}$.

Let $m \in \text{Pur}$ and let $m = \int_0^\infty \nu_t \, dt$ where ν is an entrance law. Define μ as in (6.20) and note

$$\begin{aligned}
\mu(f) &= Q_m[f \circ Y_{\alpha+};\ 0 < \alpha < 1;\ W(h)] \\
&= \int_{-\infty}^{\infty} Q_\nu[f \circ Y_{\alpha+};\ t < \alpha < 1 + t;\ W(h)] \, dt \\
&= Q_\nu[f \circ Y_{0+};\ W(h)],
\end{aligned}$$

where, of course, we have used $Q_\nu(\alpha \neq 0) = 0$ in the last step. Hence using (6.15)—or more precisely (6.16)—for the second equality,

$$\mu P_t(f) = Q_\nu[P_t f \circ Y_{0+};\ W(h)] = Q_\nu[f \circ Y_t^*;\ W(h)].$$

Consequently

$$\begin{aligned}
\mu U(f) &= \int_0^\infty Q_\nu[f \circ Y_t;\ W(h)] \, dt \\
&= \int_{-\infty}^{\infty} Q_\nu[f \circ Y_t;\ W(h)] \, dt = Q_m[f \circ Y_0;\ W(h)]
\end{aligned}$$

where we have used $Y_t = Y_t^*$ if $t > \alpha = 0$ a.s. Q_ν for the first equality. Therefore $m = \mu U + \rho$ where $\rho(f) = Q_m[f \circ Y_0;\ W(h)^c]$. Using the fact that $W(h)^c \in \mathcal{G}_{\alpha+}^*$ and the uniqueness assertion in (6.3) it is readily checked that $\rho \in \text{Exc}$ and that $Q_\rho(\cdot) = Q_m(\cdot, W(h)^c)$. Hence $\rho \in \text{Har}$ by the second paragraph of this proof and $m = \mu U + \rho$. This establishes the assertions in the third sentence of (6.19) when $m \in \text{Pur}$. In general $m = m_i + m_p$ and since $Q_{m_i}(W(h)) = 0$ and $\text{Inv} \subset \text{Har}$, one obtains the same conclusion for all $m \in$

Exc. In particular if $m \in \text{Har}$, then $Q_m(W(h)) = 0$, and if $Q_m(W(h)^c) = 0$, then $m \in \text{Pot}$. ∎

As the final topic in this section we turn to a description of Dis and Con in terms of Kuznetsov measures once again following [**FM86**]. We require some definitions.

(6.21) **Definition.** A *stationary time* is a function $S: W \to [-\infty, \infty]$ which satisfies $S \in \mathcal{G}$ and $t + S \circ \sigma_t = S$ for all $t \in \mathbb{R}$.

If S is only required to be $\mathcal{G}^m$ measurable in the above definition for some $m \in \text{Exc}$, then we call S an m-stationary time. It was shown in the paragraph preceding (6.7) that if S is m-stationary then $Q_m(S \in dt)$ is a multiple (possibly infinite) of Lebesgue measure. In particular $Q_m(S = t) = 0$ for each $t \in \mathbb{R}$. Of course, $Q_m(S \in dt)$ is regarded as a measure on $\mathbb{R}$ and not on $[-\infty, \infty]$; that is $Q_m(S \in dt) = Q_m(S \in dt;\ S \in \mathbb{R})$. A *stationary stopping time* is a stationary time S that is also a $(\mathcal{G}_t)$ stopping time and which satisfies $S \geq \alpha$. Requiring $S \geq \alpha$ in this definition is no real restriction since if S satisfies all of the other requirements, then $S \vee \alpha$ is a stationary stopping time. Finally a *stationary terminal time* T is a stationary stopping time that satisfies, in addition, $T = \infty$ if $T > \beta$ and

$$t + T \circ \theta_t = T \quad \text{on} \quad \{\alpha < t < T\}. \tag{6.22}$$

We emphasize that a stationary terminal time is defined on W while a terminal time (defined in §4) is defined on Ω. If $(\mathcal{G}_t)$ is replaced by $(\mathcal{G}^0_{t+})$ in the above definitions we shall speak of a $(\mathcal{G}^0_{t+})$ stationary stopping time (resp. terminal time). Note that (6.22) also holds on $\{-\infty < \alpha = t\}$ since on this set $\theta_t = \sigma_t$.

(6.23) **Proposition.** *Let S be a stationary stopping time and define $\mu_S(f) = Q_m[f \circ Y_S;\ 0 < S < 1]$. Then $\mu_S U(f) = Q_m[f \circ Y_0;\ S < 0;\ \alpha < S < \beta]$.*

Proof. Using the strong Markov property for the second equality we have

$$\begin{aligned}
\mu_S U(f) &= \int_0^\infty \mu_S P_t(f)\,dt \\
&= \int_0^\infty Q_m[f\circ Y_{S+t};\ 0<S<1,\ \alpha<S<\beta]\,dt \\
&= \int_{-\infty}^\infty Q_m[f\circ Y_t;\ S<t;\ 0<S<1;\ \alpha<S<\beta]\,dt \\
&= \int_{-\infty}^\infty Q_m[f\circ Y_0;\ S<0;\ -t<S<1-t,\ \alpha<S<\beta]\,dt \\
&= Q_m[f\circ Y_0;\ S<0;\ \alpha<S<\beta],
\end{aligned}$$

where the next to last equality uses the stationarity of S and of Q_m. ∎

(6.24) **Theorem.** (i) $m \in \mathrm{Con}$ *if and only if* $Q_m(S \in \mathbb{R}) = 0$ *for all stationary stopping times* S. (ii) $m \in \mathrm{Dis}$ *if and only if there exists a decreasing sequence* (S_n) *of stationary stopping times such that a.s.* Q_m *for each* n, $S_n > \alpha$, $S_n < \beta$ *on* $\{S_n < \infty\}$ *and* $S_n \downarrow\downarrow \alpha$. (iii) *If* $m \in \mathrm{Dis}$ *and* (S_n) *is a sequence as in* (ii), *then* $\mu_{S_n} U \uparrow m$ *where* μ_{S_n} *is defined as in* (6.23). (iv) *If* $m \in \mathrm{Dis}$, *then there exists a stationary time* S *with* $\alpha < S < \beta$ *a.s.* Q_m.

Proof. (i) Suppose $m \in \mathrm{Con}$ and S is a stationary stopping time. Define μ_S as in (6.23). Then $\mu_S U \le m$ and hence $\mu_S = 0$. It was pointed out in the paragraph following (6.6) that $\alpha = -\infty$ and $\beta = \infty$ a.s. Q_m when $m \in \mathrm{Con}$. It follows that $Q_m(0 < S < 1) = 0$ and then by stationarity that $Q_m(n < S < n+1) = 0$ for each integer n. Hence $Q_m(S \in \mathbb{R}) = 0$. Conversely suppose $Q_m(S \in \mathbb{R}) = 0$ for all stationary stopping times S. Let $0 < h \le 1$ with $m(h) < \infty$ and $m(\{Uh < \infty\}) > 0$. Choose $k \in p\mathcal{E}$ such that

$0 < m(kUh) < \infty$. This is possible since the measure $Uh \cdot m$ is σ-finite on $\{Uh < \infty\}$. Then

$$Q_m[h \circ Y_t \int_\alpha^t k \circ Y_s \, ds]$$

$$= \int_{-\infty}^t m(kP_{t-s}h) \, ds = m(kUh) < \infty .$$

Define $K_t = \int_\alpha^t k \circ Y_s \, ds$. Since $h > 0$, $K_t < \infty$ a.s. Q_m on $\{t < \beta\}$. Therefore (K_t) is continuous and $(\mathcal{G}_t^0)$ adapted, and since $m(kUh) > 0$, $Q_m(K_t > 0) > 0$. Define $S_n = \inf \{t\colon K_t > 1/n\}$. Then each S_n is a $(\mathcal{G}_{t+}^0)$ stationary stopping time, $S_n > \alpha$ a.s. Q_m, and for large enough n, $Q_m(\alpha < S_n < \beta) > 0$ because $Q_m(K_\beta > 0) > 0$. This contradicts our assumption, and so $Uh = \infty$ a.e. m whenever $h > 0$ with $m(h) < \infty$. Consequently $m \in \text{Con}$.

(ii) Suppose $m \in \text{Dis}$. Then there exists $0 < h \leq 1$ with $Uh < \infty$ a.e. m. Thus we may choose $k \in \mathcal{E}$ with $k > 0$ such that $m(kUh) < \infty$. Let K and S_n be defined as in the preceding paragraph. This time $0 < K_t < \infty$ a.e. Q_m on $\{\alpha < t < \beta\}$. Therefore $\alpha < S_n < \beta$ on $\{S_n < \infty\}$ and $S_n \downarrow\downarrow \alpha$ a.s. Q_m. Next suppose (S_n) is a sequence as in (ii). From (6.23)

$$\mu_{S_n} U(f) = Q_m[f \circ Y_0;\ S_n < 0] \uparrow m(f),$$

so that $m \in \text{Dis}$ and (iii) holds.

(iv) Let (S_n) be as in (ii). Define

$$\begin{aligned} S &= S_n \quad \text{on} \quad \{S_n < \infty,\ S_{n-1} = \infty\},\ n \geq 1 \\ &= \infty \quad \text{on} \quad \cap \{S_n = \infty\} \end{aligned}$$

where we set $S_0 = \infty$. Since $S_n > \alpha$ and $S_n \downarrow\downarrow \alpha$ a.s. Q_m it follows that $\alpha < S < \beta$ a.s. Q_m. It is easily checked that S is a stationary time. ∎

Remark. The above proof shows that the S_n in (ii) may be chosen to be $(\mathcal{G}^0_{t+})$ stationary stopping times and that S in (iv) may be chosen to be $\mathcal{G}^0$ measurable.

The next result sharpens the necessary and sufficient conditions in (6.24).

(6.25) **Theorem.** (i) $m \in \mathrm{Con}$ *if and only if* $Q_m(S \in \mathbb{R}) = 0$ *for all stationary times* S. (ii) $m \in \mathrm{Dis}$ *if and only if there exists a stationary time* S *with* $\alpha < S < \beta$ *a.s.* Q_m.

In order to prove this we need a preliminary result which is of considerable interest in its own right and which will be used in later sections. In order to formulate it we need some notation. Let $\mathcal{I}^m$ (resp. $\mathcal{I}$) be the σ-algebra of all sets A in $\mathcal{G}^m$ (resp. $\mathcal{G}$) that are *invariant*; that is $\sigma_t^{-1}(A) = A$ for each $t \in \mathbb{R}$. If $H \in p\mathcal{G}^m$, define $\bar{H} := \int^* H \circ \sigma_t \, dt$ where $\int^*$ is the outer integral discussed in Appendix B. Then (B.11) and the ensuing discussion shows that $\bar{H} \in \mathcal{I}^m$. Moreover a.s. Q_m, $t \to H \circ \sigma_t$ is Lebesgue measurable and $\bar{H} = \int H \circ \sigma_t \, dt$. If $H \in p\mathcal{G}$, then $\bar{H} \in \mathcal{I}$. Note that if S is a stationary time and $H := 1_{]a,b[}(S)$, then using (B.12)

$$\bar{H} = \int^* 1_{]a,b[}(S - t)\, dt = (b-a)\, 1_{\{S \in \mathbb{R}\}}\,. \tag{6.26}$$

(6.27) **Proposition.** *Let* $G, H \in p\mathcal{G}$ *and* $A \in \mathcal{I}$. *If* $m \in \mathrm{Exc}$, *then* $Q_m(H\bar{G}; A) = Q_m(\bar{H}G; A)$.

Proof. Because $\bar{G} = \int_{-\infty}^{\infty} G \circ \sigma_t \, dt$ a.s. Q_m and similarly for $\bar{H}$, Fubini's theorem and the stationarity of Q_m imply

$$
\begin{aligned}
Q_m\left(H \int_{-\infty}^{\infty} G \circ \sigma_t \, dt;\, A\right)
&= \int_{-\infty}^{\infty} Q_m(H G \circ \sigma_t;\, A)\, dt \\
&= \int_{-\infty}^{\infty} Q_m(H \circ \sigma_{-t} G;\, A)\, dt = Q_m(\bar{H} G;\, A)\,. \qquad \blacksquare
\end{aligned}
$$

(6.28) **Remarks.** We have stated (6.27) so that it applies to all $m \in \text{Exc}$ simultaneously. Of course, if $m \in \text{Exc}$ is fixed it suffices to suppose that $G, H \in p\mathcal{G}^m$ and $A \in \mathcal{G}^m$ satisfies $\sigma_t^{-1}(A) = A$ a.s. Q_m for each $t \in \mathbb{R}$. We shall sometimes use (6.27) under these (apparently) more general hypotheses.

We now prove (6.25). Because of Theorem 6.24, for (i) it suffices to show that if $m \in \text{Con}$ and S is a stationary time, then $Q_m(S \in \mathbb{R}) = 0$. Suppose there exists a stationary time S with $Q_m(S \in \mathbb{R}) > 0$. Let $0 < h \leq 1$ with $h \in \mathcal{E}$ and $m(h) < \infty$. Then there exist $-\infty < a < b < \infty$ with $Q_m(a < S < b) > 0$. Using (6.26) and (6.27)

$$Q_m\left[\int_{-\infty}^{\infty} h \circ Y_t\, dt;\ a < S < b\right] = (b-a)\, Q_m[h \circ Y_0;\ S \in \mathbb{R}]$$
$$\leq (b-a)m(h) < \infty\,,$$

and $Q_m[h \circ Y_0;\ S \in \mathbb{R}] > 0$. Therefore

$$Q_m\left[0 < \int_{-\infty}^{\infty} h \circ Y_t\, dt < \infty\right] > 0\,.$$

Defining $S_n = \inf\left\{t\colon \int_{-\infty}^{t} h \circ Y_s\, ds > 1/n\right\}$, it follows that for n large enough $Q_m(S_n \in \mathbb{R}) > 0$. But S_n is a stationary stopping time contradicting $m \in \text{Con}$. For (ii) we must show that if S is a stationary time with $\alpha < S < \beta$ a.s. Q_m, then $m \in \text{Dis}$. Choosing h as above, it follows from (6.27) as before, that $\int_{-\infty}^{\infty} h \circ Y_t\, dt < \infty$ a.s. Q_m on $\{a < S < b\}$ for any $a < b$ and hence on $\{S \in \mathbb{R}\} = W$ a.s. Q_m. Defining S_n as above the sequence (S_n) has the properties in (6.24ii), and so $m \in \text{Dis}$. ∎

We leave the following two assertions to the reader as exercises. (Use (3.9) and the fact that $Q_m(\alpha \in dt) = Q_m(0 < \alpha < 1)\, dt$ for the first.)

(6.29) (i) $Q_m(W) < \infty \Longrightarrow m \in \text{Con}$.

(ii) $\xi \prec \eta \Longleftrightarrow Q_\xi \leq Q_\eta$ for $\xi, \eta \in \text{Exc}$.

7. Kuznetsov Measures II

In this section we shall study the relationship between Kuznetsov measures and the energy functional and the balayage operation. As in §6, throughout this section we suppose that (6.2) is in force. The right process X is always realized on the Ω defined below (6.5). We begin by expressing the energy functional in terms of the Kuznetsov measure.

Let $H \in p\mathcal{F}$ so that H is defined on Ω. We say that H is *excessive* provided $s \to H \circ \theta_s$ is decreasing and right continuous on $[0, \infty[$. Then $h(x) := P^x(H)$ is an excessive function for X. If $\alpha(w) < t$, then $\theta_t w \in \Omega$ and if $\alpha < s < t$, $H \circ \theta_t = H \circ \theta_{t-s} \circ \theta_s \leq H \circ \theta_s$. Therefore we extend H to all of W by

$$(7.1) \qquad H^* = \uparrow \lim_{t \downarrow \alpha} H \circ \theta_t = \sup_{t > \alpha} H \circ \theta_t = \sup_{r \in \mathbb{Q}, r > \alpha} H \circ \theta_r .$$

Clearly H^* is $\mathcal{G}$ measurable. If $H \in \mathcal{F}^*$ then $H^* \in \mathcal{G}^*$. It is evident that H^* is invariant; that is $H^* \circ \sigma_t = H^*$ for all t. The next result comes from **[F88a]**.

(7.2) **Proposition.** *Let H, h, and H^* be as above. Let $m \in$ Dis and S be a stationary time with $\alpha < S < \beta$ a.s. Q_m. Then $L(m, h) = Q_m(H^*;\ 0 < S < 1)$.*

Proof. Since H^* is invariant, it follows from (6.26) and (6.27) that the right hand side of the formula in (7.2) is the same for all stationary times S such that $\alpha < S < \beta$ a.s. Q_m. Let (S_n) be a sequence of stationary stopping times as in (6.24-ii). Recalling the definition of $\mu_n := \mu_{S_n}$ in (6.23),

$$\begin{aligned} \mu_n(h) &= Q_m[h \circ Y_{S_n};\ 0 < S_n < 1] \\ &= Q_m[H \circ \theta_{S_n};\ 0 < S_n < 1,\ S \in \mathbb{R}] \\ &= Q_m[H \circ \theta_{S_n};\ 0 < S < 1,\ S_n \in \mathbb{R}], \end{aligned}$$

where for the second equality we have used the strong Markov property, $S_n > \alpha$, and $S \in \mathbb{R}$ a.s. Q_m, while for the third we have used the invariance of $H \circ \theta_{S_n}$ and (6.27). Now let $n \to \infty$. By (6.24-iii), $\mu_n U \uparrow m$ and so $\mu_n(h) \uparrow L(m,h)$, while a.s. Q_m, $S_n \downarrow\downarrow \alpha$ so that $H \circ \theta_{S_n} \uparrow H^*$ and $\{S_n \in \mathbb{R}\} \uparrow W$. ∎

This expresses $L(m,h)$ as an expectation relative to Q_m when h is of the form $h = P^{\cdot}(H)$ with H excessive. This seems to be the most useful such formula. We refer the reader to [**GSt87**, (5.14)] for other formulas for the energy functional in terms of the Kuznetsov measure.

We now define for S a stationary time and $\xi \in \text{Exc}$

$$r_S\xi(f) = Q_\xi[f \circ Y_t;\ S < t]. \tag{7.3}$$

Clearly $r_S\xi$ does not depend on the choice of t in (7.3) and $r_S\xi$ is a measure with $r_S\xi \leq \xi$. If S is a stationary stopping time, then, using the Markov property for $t > 0$

$$\begin{aligned} r_S\xi(P_t f) &= Q_\xi(P_t f \circ Y_0;\ S < 0) = Q_\xi(f \circ Y_t;\ S < 0;\ \alpha < 0) \\ &= Q_\xi(f \circ Y_0;\ S < -t;\ \alpha < -t) \end{aligned}$$

which increases to $r_S\xi(f)$ as $t \downarrow 0$. Hence $r_S\xi \in \text{Exc}$. This should be compared to the situation over X: If $u \in S(X)$ and T is a stopping time, then $P_T u \leq u$, while if T is a terminal time (resp. exact terminal time), $P_T u$ is supermedian (resp. $P_T u \in S(X)$). *Let us write SST for the class of stationary stopping times.* Then if $S \in SST$, $r_S\colon \text{Exc} \to \text{Exc}$, $r_S(a\xi + b\eta) = a r_S(\xi) + b r_S(\eta)$ for $a, b \geq 0$ and $\xi, \eta \in \text{Exc}$, and $r_S\xi \leq \xi$.

(7.4) **Proposition**. *Let $S \in SST$ and $\xi \in \text{Exc}$.*

(i) $(r_S\xi)_c = r_S\xi_c$, $(r_S\xi)_d = r_S\xi_d$.

(ii) *If $(\xi_n) \subset \text{Exc}$ increases to ξ in the strong order, then $(r_S\xi_n)$ increases to $r_S\xi$ in the strong order.*

(iii) *If $(\xi_n) \subset \text{Exc}$ and $\xi_n \uparrow \xi$, then $r_S\xi_{nc}$ increases to $r_S\xi_c$ in the strong order.*

Proof. Since $r_S\xi_c \le \xi_c$ and $r_S\xi_d \le \xi_d$ it follows that $r_S\xi_c \in$ Con and $r_S\xi_d \in$ Dis. Since $r_S\xi = r_S\xi_c + r_S\xi_d$, (i) follows from the uniqueness of the decomposition (2.4). If (ξ_n) increases to ξ in the strong order, then defining $\eta_n = \xi_n - \xi_{n-1}$ for $n \ge 1$ where $\xi_0 = 0$, one has $\eta_n \in$ Exc and $\xi = \Sigma_n \eta_n$. The uniqueness in Theorem 6.3 then implies that $Q_\xi = \Sigma_n Q_{\eta_n}$. Clearly this yields (ii). If $\xi_n \uparrow \xi$, then $\xi_{nc} \uparrow \xi_c$ and $\xi_{nd} \uparrow \xi_d$ because of (2.4). Now (iii) follows from (ii) since the simple and strong orders agree on Con. ∎

Recall the definition of the birthing operators b_t defined in (6.5) and the sets $W(h)$ defined above (6.12). The next result and also (7.4) are lifted from **[FM86]**.

(7.5) **Proposition.** *Let $S \in SST$ and $m \in$ Exc. Then:*

(i) $Q_{r_S m}(F) = Q_m(F \circ b_S;\ S < \beta)$;

(ii) $(r_S m)_p = Q_m(Y_t \in \cdot\,;\ -\infty < S < t) = \int_0^\infty \nu_t\, dt$ *where* $\nu_t = Q_m(Y_{S+t} \in \cdot\,;\ 0 < S < 1),\ t > 0$.

(iii) *The potential part of $r_S m$ is $\mu_S U$ where*

$$\mu_S = Q_m(Y_{S+} \in \cdot\,;\ 0 < S < 1\,; W(h) \circ b_S)$$

and $0 < h \le 1$ satisfies $m(h) < \infty$. We have written $W(h) \circ b_S$ in place of $b_S^{-1}(W(h))$.

Proof. Recall that for a stationary stopping time S one has $S \ge \alpha$. If $F = \prod_{j=1}^n f_j \circ Y_{t_j}$, $t_1 < \cdots < t_n$, then $1_{\{S<\beta\}} F \circ b_S = 1_{\{S<t_1\}} F$ since $f_j(\Delta) = 0$ by convention. It is now easy to check that (i) holds for such F. Hence (i) follows because $\beta([\Delta]) = -\infty$ and a Kuznetsov measure is uniquely determined by its finite dimensional distributions and the fact that it does not charge $[\Delta]$. Since $\alpha \circ b_S = S$ if $S < \beta$, (ii) follows from (i) and (6.7), while (iii) follows from (i) and (6.19). ∎

We are next going to relate the operators r_S to the operators R_T defined in §4. However, before coming to this we shall introduce a technical simplification. Let T be an exact terminal time over X as defined in §4. We suppose that $T([\Delta]) = \infty$. Since Ω is now (isomorphic to) the canonical space for X, it follows from (A.38) that there exists an exact terminal time S with $S = T$ a.s. and such that S is an $(\mathcal{F}^*_{t+})$ stopping time. Because both S and T are exact, it is clear that the processes $(t + S \circ \theta_t)_{t \geq 0}$ and $(t + T \circ \theta_t)_{t \geq 0}$ are indistinguishable. Of course, $P^q_T(x, \cdot) = P^q_S(x, \cdot)$ for each $q \geq 0$ and $x \in E$, and $R^q_T \xi = R^q_S \xi$ for each $q \geq 0$ and $\xi \in \mathrm{Exc}$. In view of these facts we shall replace T by S in the sequel. Therefore in the remainder of this monogrph the following convention is in force (unless explicitly stated otherwise):

(7.6) *All terminal times, T, over X are assumed to be $(\mathcal{F}^*_{t+})$ stopping times and are normalized by* $T([\Delta]) = \infty$.

By (A.36) and (A.37) any such T must satisfy $T = \infty$ on $\{T > \zeta\}$ and $T \circ k_t = T$ on $\{T < t\}$. If T is an exact terminal time such that $T = \infty$ on $\{T \geq \zeta\}$ (this is (A.34ii)) and $T \circ k_t = \infty$ on $\{T \geq t\}$, then the S from (A.38) is such that the processes $(S \circ k_t)_{t \geq 0}$ and $(T \circ k_t)_{t \geq 0}$ are also indistinguishable. See (A.38). In particular this applies to T_B when $B \in \mathcal{E}^e$. See the discussion in the last paragraph of Appendix A. We stress that (7.6) applies to *all* terminal times and not just to exact terminal times.

A word of caution is in order. Given $B \in \mathcal{E}^e$ there exists an exact terminal time, T, satisfying (7.6) and such that (i)$T \geq \zeta \Rightarrow T = \infty$, (ii) $T < t \Rightarrow T \circ k_t = T$, (iii) $T \geq t \Rightarrow T \circ k_t = \infty$, (iv) the processes $(T \circ \theta_t)$ and $(T_B \circ \theta_t)$ are indistinguishable, and (v) the processes $(T \circ k_t)$ and $(T_B \circ k_t)$ are indistinguishable. *From now on we write T_B for such a version T of* $\inf\{t > 0: X_t \in B\}$. Of course, if $B \in \mathcal{E}$, $T_B(\omega) = \inf\{t > 0: X_t(\omega) \in B\}$ for every ω has all of these properties. However when $B \in \mathcal{E}^e$ but not in $\mathcal{E}$ it may be necessary to modify $\inf\{t > 0: X_t \in B\}$ to obtain an $(\mathcal{F}^*_{t+})$ stopping time. If $B \in \mathcal{E}^*$, then by (B.10), σ_B, the Lebesgue penetration time of B, is already an $(\mathcal{F}_{t+}^*)$ stopping time without modification and it clearly satisfies (i), (ii), and (iii) just above as well.

Let T be a terminal time for X. Then T is defined on Ω and by our convention satisfies the conditions in (7.6). If $\alpha(w) < t$, then $\theta_t w \in \Omega$, and so for $\alpha(w) < s < t$

$$t + T(\theta_t w) = s + (t - s) + T(\theta_{t-s}\theta_s w) \geq s + T(\theta_s w).$$

Therefore we may define on all of W

$$(7.7) \qquad \tau := \downarrow \lim_{t \downarrow \alpha} (t + T \circ \theta_t) = \inf_{r > \alpha, r \in \mathbb{Q}} (r + T \circ \theta_r)$$

with $\tau([\Delta]) = \infty$ by convention. Since θ_t as a map from $\{\alpha < t\}$ to Ω is measurable relative to $\mathcal{G}_{s+t}^*\big|_{\{\alpha<t\}}$ and $\mathcal{F}_s^*$ for each $s \geq 0$, it follows that τ is $(\mathcal{G}_{t+}^*)$ stopping time. Note that if $\alpha < s < \beta < \tau$, then $T \circ \theta_s \geq \tau - s > \beta - s = \zeta \circ \theta_s$ and, hence, $T \circ \theta_s = \infty$. Therefore $\tau = \infty$ if $\tau > \beta$. Obviously $\tau \geq \alpha$ and τ is stationary because $\theta_t \sigma_s = \theta_{t+s}$ for $s, t \in \mathbb{R}$. We next claim that τ has the terminal time property (6.22). To this end first observe that if $\alpha < s < t < \tau$, then $t - s < \tau - s \leq T \circ \theta_s$ and so

$$t + T \circ \theta_t = s + (t - s) + T \circ \theta_{t-s} \circ \theta_s = s + T \circ \theta_s;$$

that is, $t \to t + T \circ \theta_t$ is constant and equal to τ on $]\alpha, \tau[$. We now fix w but suppress it (for the most part) in our notation. We claim that $\tau(\theta_t) = \tau - t$ if $t < \tau$. This is obvious if $w = [\Delta]$ or if $t \leq \alpha$ since $t \leq \alpha \Longrightarrow \theta_t = \sigma_t$. Next suppose that $\alpha < t < \beta$ so that $\alpha(\theta_t) = 0$. Then

$$\begin{aligned} \tau(\theta_t) &= \lim_{s \downarrow \alpha(\theta_t)} [s + T(\theta_s \circ \theta_t)] \\ &= \lim_{s \downarrow t} [s + T(\theta_s)] - t = \tau - t, \end{aligned}$$

because $s + T(\theta_s) = \tau$ for $t < s < \tau$. Finally suppose $t < \tau$ and $\beta \leq t$. Then $\theta_t w = [\Delta]$ and so $\tau(\theta_t) = \infty$. But $\tau > \beta$

and so $\tau = \infty$. Consequently τ is a stationary terminal time as defined in §6 and τ is also a $(\mathcal{G}^*_{t+})$ stopping time. We also claim that $\tau \circ k_t = \tau$ on $\{\tau < t\}$. If $t > \tau$, then $t - s > T(\theta_s)$ for s close enough to α. Also $\alpha(k_t) = \alpha$ if $\alpha < t < \beta$, and so using the fact that $T \circ k_u = T$ on $\{T < u\}$ for the last equality

$$\tau(k_t) = \lim_{s \downarrow \alpha} [s + T(\theta_s \circ k_t)] = \lim_{s \downarrow \alpha} [s + T(k_{t-s} \circ \theta_s)] = \tau .$$

If $T = T_B$, the hitting time of B by X, then τ defined by (7.7) is the hitting time of B by Y; that is, $\tau = \inf\{t\colon Y_t \in B\}$. In view of the word of caution following (7.6), if $B \in \mathcal{E}^e$ but not in $\mathcal{E}$, one can only assert that $\tau = \inf\{t\colon Y_t \in B\}$ a.s. Q_m for each $m \in \text{Exc}$ and a.s. Q_ν for each entrance rule ν. We adopt the same convention for Y as for X; that is, τ_B will denote a version of $\inf\{t\colon Y_t \in B\}$ coming from (7.7) and a good version of T_B. Thus τ_B is a $(\mathcal{G}^*_{t+})$ stopping time for all $B \in \mathcal{E}^e$. If $B \in \mathcal{E}^*$ and σ_B is the Lebesgue penetration time of B by X, then $\tau := \downarrow \lim_{t \downarrow \alpha} (t + \sigma_B \circ \theta_t)$ is the Lebesgue penetration time of B by Y; that is

$$\tau = \inf\left\{t > \alpha\colon \int_\alpha^t 1_B \circ Y_s \, ds > 0\right\} .$$

(Since $\mathcal{G}^0$ contains all singletons, $\{w\}$, it is not necessary to use the outer integral here. See Appendix B.)

If T is a terminal time and $\tilde{T}$ is its exact regularization then $\tau = \tilde{\tau}$ where τ and $\tilde{\tau}$ are defined from T and $\tilde{T}$ as in (7.7). Thus the extension of T to W embodied in (7.7) does not distinguish between T and its exact regularization. In particular the map $T \to \tau$ is not injective. Nonetheless we shall simplify our notation and denote the extension of a terminal time T defined on Ω to all of W using the recipe (7.7) by the same letter T rather than τ. To repeat, $T = \downarrow \lim_{t \downarrow \alpha} (t + T \circ \theta_t)$ on W. Note that with T so defined on W its restriction to Ω is the exact regularization of the original

T on Ω. So, at least for exact terminal times, the notation is consistent. In particular we shall use T_B and σ_B to denote the hitting time of B and the Lebesgue penetration time of B respectively by either X or Y.

Reflecting (7.6) we adopt the following conventions for stationary times over Y (see (6.21)) unless explicitly stated otherwise:

(7.8) (i) *All stationary times are assumed to be* $(\mathcal{G}^*)$ *measurable.*

(ii) *A stationary stopping time,* S, *is a stationary time that is a* $(\mathcal{G}^*_{t+})$ *stopping time and satisfies* $S \geq \alpha$.

(iii) *A stationary terminal time is a stationary stopping time* S *satisfying* (6.22), $S = \infty$ *if* $S > \beta$, *and* $S \circ k_t = S$ *if* $S < t$.

The conventions (7.6) and (7.8) are consistent in that if T is a terminal time over X, then its extension T to W using (7.7) is a stationary terminal time. We shall use the abbreviations ST, SST, and STT for the classes of stationary times, stationary stopping times, and stationary terminal times respectively.

(7.9) **Theorem.** *Let* T *be an exact terminal time for* X. *Then* $r_T m = R_T m$ *for each* $m \in \text{Exc}$.

Proof. Suppose first that $m \in \text{Dis}$. Given $f \in p\mathcal{E}$ define $H = \int_T^\infty f \circ X_s \, ds$ on Ω. Since T is exact $s \to H \circ \theta_s$ is decreasing and right continuous on $[0, \infty[$. It is evident that H^* defined in (7.1) is given by $H^* = \int_T^\infty f \circ Y_t \, dt$ on W. Since $P^x(H) = P_T U f(x)$, we obtain from (4.12) and (7.2)

$$R_T m(f) = L(m, P_T U f) = Q_m(H^*;\ 0 < S < 1)$$

where S is a stationary time with $\alpha < S < \beta$ a.s. Q_m. But $H^* = \int_{-\infty}^{\infty} (f \circ Y_0 1_{\{T<0\}}) \circ \sigma_t \, dt$. Consequently from (6.26) and (6.27), $R_T m(f) = Q_m(f \circ Y_0;\, T < 0) = r_T m(f)$ because $\alpha < S < \beta$ a.s. Q_m. This establishes (7.9) for $m \in \text{Dis}$.

Next suppose $m \in \text{Con}$. Then $\alpha = -\infty$ and $\beta = \infty$ a.s. Q_m, and using (4.17)—$\phi(x) = P^x(T < \infty)$ as in (4.16)—

$$R_T m(f) = m(\phi f) = Q_m(f \circ Y_0;\ T \circ \theta_0 < \infty).$$

Define $S = \sup\{t\colon t + T \circ \theta_t < \infty\}$ where the supremum of the empty set is minus infinity. Because $\theta_t \sigma_s = \theta_{t+s}$, it is clear that S is a stationary time. Since $T = \downarrow \lim_{t \to -\infty} (t + T \circ \theta_t)$, $\{S > -\infty\} = \{T < \infty\}$. Moreover $m \in \text{Con}$ implies that $Q_m(S \in \mathbb{R}) = 0 = Q_m(T \in \mathbb{R})$. Consequently a.s. Q_m,

$$\{T \circ \theta_0 < 0\} = \{S \geq 0\} = \{S > -\infty\} = \{T < \infty\} = \{T < 0\},$$

and so $R_T m(f) = Q_m(f \circ Y_0;\ T < 0) = r_T m(f)$. ∎

We now are prepared to give the definitive form of Fitzsimmons' theorem (5.21), albeit under the additional hypothesis (6.2).

(7.10) **Theorem.** *Let ξ and m be excessive measures. Then, under (6.2), there exists an increasing family $\{S(t);\, t \geq 0\}$ of stationary terminal times (defined on W) such that*

$$\xi = \int_0^\infty r_{S(t)} m\, dt + r_S \xi, \tag{7.11}$$

where $S = \uparrow \lim_{t \to \infty} S(t)$. If $\xi \leq am$ for some $a < \infty$, then $\xi = \int_0^a r_{S(t)} m\, dt$.

Proof. Recall the notation of Theorem 5.21 and the definition of $A(t, \epsilon)$ in (5.15). Set $\sigma(t, \epsilon) = \sigma_{A(t,\epsilon)}$. Then as remarked earlier, the extension of $\sigma(t, \epsilon)$ to W is given by

$$\sigma(t, \epsilon) = \inf\left\{s > \alpha\colon \int_\alpha^s 1_{A(t,\epsilon)} \circ Y_u\, du > 0\right\},$$

and using (7.9), we obtain from (5.17),

$$\delta_t = \downarrow \lim_{\epsilon \downarrow 0} r_{\sigma(t,\epsilon)} m \,. \tag{7.12}$$

Since $A(t,\epsilon)$ decreases as ϵ decreases we may define on W

$$S(t) = \uparrow \lim_{\epsilon \downarrow 0} \sigma(t,\epsilon); \quad t \geq 0 \,, \tag{7.13}$$

and each $S(t)$ is easily seen to be a stationary terminal time. Because $A(t,\epsilon)$ decreases as t increases, it is clear that $\{S(t); t \geq 0\}$ is an increasing family of stationary terminal times. Recall that if $\eta \in \text{Exc}$ and S is any stationary time, then $Q_\eta\,(S=0)=0$. Using this, (7.3), (7.13), and (7.9),

$$r_{S(t)}\eta = \downarrow \lim_{\epsilon \downarrow 0} r_{\sigma(t,\epsilon)}\eta = \downarrow \lim_{\epsilon \downarrow 0} R_{\sigma(t,\epsilon)}\eta \,. \tag{7.14}$$

It follows from (7.14) that if $\eta_1, \eta_2 \in \text{Exc}$ with $\eta_1 \leq \eta_2$, then $r_{S(t)}\eta_1 \leq r_{S(t)}\eta_2$. Combining (7.14), (7.12), and (5.22) we obtain

$$\xi = \gamma_\infty + \int_0^\infty r_{S(t)} m \, dt$$

where $\gamma_\infty \downarrow \lim \gamma_t$ and γ_t is defined in (5.14).

Thus to complete the proof of (7.10) we must identify γ_∞ as $r_S\xi$. To this end first note that $S := \uparrow \lim_{t\to\infty} S(t)$ is a stationary terminal time and that $r_S\eta = \downarrow \lim_{t\to\infty} r_{S(t)}\eta$ for any $\eta \in \text{Exc}$. In particular $r_{S(t)}m \downarrow r_S m$. But if $f > 0$ with $\xi(f) < \infty$, then $\int_0^\infty r_{S(t)}m(f)\,dt \leq \xi(f) < \infty$ and so $r_{S(t)}m \downarrow 0$ as $t \to \infty$. Hence $r_S m = 0$. By (5.16) and (7.9), $\gamma_t = r_{\sigma(t,\epsilon)}\gamma_t$. Letting $\epsilon \downarrow 0$ we obtain $\gamma_t = r_{S(t)}\gamma_t \leq r_{S(t)}\gamma_u$ if $0 \leq u < t$ since $\gamma_t \leq \gamma_u$. Now let $t \to \infty$ to obtain $\gamma_\infty \leq r_S\gamma_u$ for each $u \geq 0$. Rewrite (5.20) in the form $\xi = \gamma_t + \int_0^t r_{S(u)}m\,du$. Since the integral is an excessive measure that is dominated by

tm and $r_S m = 0$, we obtain $r_S\xi = r_S\gamma_t \leq \gamma_t$. Hence $\gamma_\infty \leq r_S\gamma_t = r_S\xi \leq \gamma_t$ and letting $t \to \infty$ we obtain $r_S\xi = \gamma_\infty$. ■

(7.15) **Corollary.** *If $\xi, m \in$ Exc with $\xi \leq m$ and T is a stationary terminal time, then $r_T\xi \leq r_T m$.*

Proof. By (7.10), $\xi = \int_0^1 r_{S(t)} m \, dt$ where $\{S(t);\, 0 \leq t \leq 1\}$ is an increasing family of stationary terminal times. By checking finite dimensional distributions one sees that $Q_\xi = \int_0^1 Q_{r_{S(t)}m}\, dt$. Hence using (7.5i)

$$r_T\xi(f) = \int_0^1 Q_m[f \circ Y_0;\; T \circ b_{S(t)} < 0;\; S(t) < 0]\, dt .$$

Now $b_u = \sigma_{-u}\theta_u$ and so $T \circ b_u = u + T \circ \theta_u \geq T$ because T is a stationary terminal time. Therefore $r_T\xi(f) \leq r_T m(f)$. ■

(7.16) **Remarks.** We shall say that a stationary terminal time S is *exact* provided $S = \downarrow \lim_{t \downarrow \alpha} (t + S \circ \theta_t)$. If S is exact and T is its restriction to Ω, then T is an exact terminal time for X and the extension of T to W by (7.7) gives S back again. Hence $r_S\eta = R_T\eta$ for $\eta \in$ Exc. If S is only a stationary terminal time and T is the restriction of S to Ω, then τ defined in (7.7) is the "exact regularization" of S. It should be emphasized that the $S(t)$ in (7.10) are not, in general, exact and so they are not the extensions of the $T(t)$ defined in (5.18) to W.

Suppose $(\xi_n) \subset$ Exc and $\xi_n \uparrow \xi \in$ Exc. If T is a stationary terminal time then according (7.15), $r_T\xi_n \uparrow \eta \leq r_T\xi$ and, of course, $\eta \in$ Exc. It is natural to ask if $\eta = r_T\xi$. This is the case if T is exact, since then $r_T\xi_n = R_T\xi_n \uparrow R_T\xi = r_T\xi$ where in R_T, T denotes the restriction of T to Ω. Also (7.4iii) implies that this is true, even for stationary stopping times T, when $\xi \in$ Con. Here is an example to show that it is not always true. Let $E = [0, \infty[$ and $P_t(x, \cdot) = \epsilon_{x+t}$, $x \geq 0$, $t \geq 0$ be the semigroup of translation to the right at unit speed. Let $\xi_n = U\left(\frac{1}{n}, \cdot\right)$ and $\xi = U(0, \cdot)$. Then $\xi_n \uparrow \xi$. Let $T = \inf\{t \geq \alpha: Y_{t+} = 0\}$. It is evident that T is a stationary terminal time and using (6.19) it follows that $r_T\xi_n = 0$ for each n while $r_T\xi = \xi$.

We next give an amusing extension of (5.23). Let $\xi \in \text{Exc}$ and $\mu U \in \text{Pot}$. Applying (5.21) to ξ and $m:= \mu U$ and arguing as in the proof of (5.23) we obtain

$$(7.17) \qquad \xi = \gamma_\infty + \nu U \quad \text{where} \quad \nu = \int_0^\infty \mu P_{T(t)}\, dt\,,$$

and the increasing family $\{T(t);\, t \geq 0\}$ of terminal times (over X) is defined in (5.18). In the next proposition "ess sup" is the essential supremum relative to Lebesgue measure on $\mathbb{R}$.

(7.18) **Proposition.** *Let $\xi \in \text{Exc}$, $\mu U \in \text{Pot}$ with $\xi << \mu U$. Let ψ be a Borel measurable version of $d\xi/d(\mu U)$. Then $\xi = \nu U$ with ν as in (7.17) if and only if for each $t \in \mathbb{R}$, $\operatorname{ess\,sup}_{s \leq t} \psi \circ Y_s < \infty$ a.s. Q_ξ on $\{t < \beta\}$.*

Proof. In view of (7.17) we must show that the condition in (7.18) is equivalent to $\gamma_\infty = 0$. From (5.26), $\gamma_\infty = \downarrow \lim_{t\to\infty} R^*_{\{\psi > t\}}\xi$. Let $\sigma(t) = \sigma_{\{\psi>t\}}$ be the Lebesgue penetration time of $\{\psi > t\}$. Then because of (4.26), $\gamma_\infty = 0$ if and only if $\downarrow \lim_{t\to\infty} R_{\sigma(t)}\xi = 0$. Clearly $\sigma(t) \uparrow \sigma(\infty)$. Let $f > 0$ with $\xi(f) < \infty$. Now (7.9) implies that

$$\lim_{t\to\infty} R_{\sigma(t)}\xi(f) = Q_\xi(f \circ Y_0;\, \sigma(\infty) < 0)\,,$$

and so $\gamma_\infty = 0$ if and only if $Q_\xi(\sigma(\infty) < 0 < \beta) = 0$. But $\sigma(\infty)$ is a stationary time and consequently $\gamma_\infty = 0$ is equivalent to $\sigma(\infty) \geq \beta$ a.s. Q_ξ. Using the fact that $\sigma(t)$ is the Lebesgue penetration time of $\{\psi > t\}$ by Y, one readily checks that $\sigma(\infty) \geq \beta$ a.s. Q_ξ is equivalent to the condition in (7.18). ∎

Remarks. If $\xi \leq k(\mu U)$ where $k < \infty$, then the condition in (7.18) is satisfied and we recover (5.23)—which we only wrote when $k = 1$ for simplicity. A similar argument shows that if $\xi = \lambda U$ with $\xi << \mu U$, then $\lambda = \nu$ with ν as in (7.17) if and only if for each $t > 0$, $\operatorname{ess\,sup}_{s \le t} \psi \circ X_s < \infty$ a.s. P^λ on $\{t < \zeta\}$.

8. Homogeneous Random Measures

For the first part of this section we do *not* suppose that condition (6.2) holds. Thus X is an arbitrary right process as in §1. In the following definition $\mathbb{R}^{++} :=]0, \infty[$ and $\mathcal{B}^{++}$ denotes the σ-algebra of Borel subsets of $\mathbb{R}^{++}$.

(8.1) **Definition.** A *random measure* (RM), $N = (N(\omega, dt))$ is a kernel from $(\Omega, \mathcal{F}^*)$ to $(\mathbb{R}^{++}, \mathcal{B}^{++})$ such that: (i) for every ω, $N(\omega, \cdot)$ is carried by $]0, \zeta(\omega)]$, and (ii) there exist bounded kernels $(N_k)_{k \geq 1}$ from $(\Omega, \mathcal{F}^*)$ to $(\mathbb{R}^{++}, \mathcal{B}^{++})$ such that $N = \sum_k N_k$.

The boundedness of N_k means that $\sup_{\omega \in \Omega} N_k(\omega, \mathbb{R}^{++}) < \infty$. If one only assumes that each N is a finite kernel, that is, $a_k(\omega) := N_k(\omega, \mathbb{R}^{++}) < \infty$ for each ω, then

$$N_{k,\ell}(\omega, \cdot) := 1_{[\ell-1, \ell[}(a_k(\omega))\, N_k(\omega, \cdot), \qquad \ell = 1, 2, \ldots;\ k = 1, 2, \ldots$$

are bounded kernels from $(\Omega, \mathcal{F}^*)$ to $(\mathbb{R}^{++}, \mathcal{B}^{++})$ whose sum is N. Consequently one may replace "bounded" by the apparently weaker "finite" in (8.1). Of course, this is also equivalent to saying that N is the limit of an increasing sequence of bounded (or finite) kernels from $(\Omega, \mathcal{F}^*)$ to $(\mathbb{R}^{++}, \mathcal{B}^{++})$. The reader should note that this definition differs somewhat from that in **[S]**, but it is more convenient for us. In particular for each ω, $N(\omega, \cdot)$ is s-finite (that is, a countable sum of finite measures) and this enables us to use the Fubini theorem. Note that $N([\Delta], \cdot) = 0$ because $\zeta([\Delta]) = 0$.

Let N be a bounded RM and $Z \in pb(\mathcal{B}^{++} \otimes \mathcal{F}^0)^*$. Of course, $(\mathcal{B}^{++} \otimes \mathcal{F}^0)^* = (\mathcal{B}^{++} \otimes \mathcal{F}^*)^*$. Given a finite measure Q on $(\Omega, \mathcal{F}^0)$ define $\tilde{Q}$ on $\mathcal{B}^{++} \otimes \mathcal{F}^0$ by

$$\tilde{Q}(F) := Q \int F(t, \cdot) N(dt) = \int Q(d\omega) \int F(t,\omega) N(\omega, dt).$$

Then $\tilde{Q}$ is a finite measure and so there exist $Z_1, Z_2 \in pb(\mathcal{B}^{++} \otimes \mathcal{F}^0)$ with $Z_1 \leq Z \leq Z_2$ and $\tilde{Q}(Z_1) = \tilde{Q}(Z_2)$. It follows that the set, Λ, of ω such that $t \to Z(t,\omega)$ is not $N(\omega, \cdot)$ measurable has Q measure zero. Since Q is arbitrary,

$$Z * N(\omega, B) := \int^* 1_B(t) Z(t,\omega) N(\omega, dt)$$

defines a bounded kernel from $(\Omega, \mathcal{F}^*)$ to $(\mathbb{R}^{++}, \mathcal{B}^{++})$ and $Z * N(\omega, B) = \int_B Z(t,\omega) N(\omega, dt)$ for every $B \in \mathcal{B}^{++}$ when $\omega \notin \Lambda$. See (B.3). This obviously extends to a general RM, N, and $Z \in p(\mathcal{B}^{++} \otimes \mathcal{F}^0)^*$, and so $Z * N$ defined as above is a RM. In particular, if $f \in p\mathcal{E}^*$ then

$$f * N(\omega, B) := \int^* 1_B(t) f \circ X_t(\omega) N(\omega, dt)$$

defines a RM. In the sequel we shall just write $Z * N(\omega, dt) = Z(t,\omega) N(\omega, dt)$; it being understood that there may be an exceptional set Λ of ω such that $Q(\Lambda) = 0$ for *every* probability Q on $(\Omega, \mathcal{F}^0)$. A similar convention holds for $f * N$ when $f \in p\mathcal{E}^*$. If $\mathcal{F}^0$ contains all singletons, then taking $Q = \epsilon_\omega$ it follows that $t \to Z(t,\omega)$ is $N(\omega, \cdot)$ measurable for every $\omega \in \Omega$ and so it is not necessary to use the upper integral in the preceding definitions. This is the case whenever (6.2) is in force since then Ω is (isomorphic to) the canonical path space for X.

An *increasing process* $A = (A_t(\omega);\ t \geq 0)$ is a process such that $A_0(\omega) = 0$, $t \to A_t(\omega)$ is increasing and right continuous on $[0, \infty[$, finite on $[0, \zeta(\omega)[$, constant on $[\zeta(\omega), \infty[$, and $A_t \in \mathcal{F}^*$ for each $t \geq 0$. If A is an increasing process, then $N(\omega, dt) := dA_t(\omega)$ is a RM. Conversely if N is a RM, then $A_t(\omega) := N(\omega,]0,t])$ is an increasing process provided it is finite for $t < \zeta(\omega)$. If N_1 and N_2 are RMs, then we write $N_1 = N_2$ provided the measures $N_1(\omega, \cdot)$ and $N_2(\omega, \cdot)$ are equal for almost all ω. (Recall almost all or almost surely means almost surely P^μ for every probability μ on E.) If N is a RM, it is often convenient to regard $N(\omega, \cdot)$ as a measure on $\mathbb{R}$ that is carried by $]0, \zeta(\omega)] \cap \mathbb{R}$, and we shall do so without special mention.

(8.2) **Definition.** A *homogeneous random measure* (HRM), N, is a RM such that $N(\theta_t\omega, B) = N(\omega, B+t)$ for all $\omega \in \Omega$, $t \geq 0$, and $B \in \mathcal{B}^{++}$.

Again this definition differs somewhat from that in [**S**]. We emphasize that it is *not* assumed that the N_k in (8.1) are homogeneous. If N is a HRM and $\varphi \in p\mathcal{B}^{++}$ then for each $s \geq 0$

$$(8.3) \qquad \int \varphi(t+s) N(\theta_s\omega, dt) = \int_{]s,\infty[} \varphi(t) N(\omega, dt)\,.$$

From now on we shall usually suppress the ω in our notation writing $N(dt)$ for $N(\cdot, dt)$ and $N(dt) \circ \theta_s$ or $N(\theta_s, dt)$ for $\omega \to N(\theta_s\omega, dt)$. Given a HRM, N, we define for $q \geq 0$ and $f \in p\mathcal{E}^*$

$$(8.4) \qquad U_N^q f(x) = P^x \int_0^\infty e^{-qt} f \circ X_t N(dt)\,.$$

Since $f(\Delta) = 0$ by convention, the integral in t in (8.4) extends only over $]0, \zeta[$. Using (8.3) it is easy to check [**S**, §36] that $U_N^q f$ is q-excessive for $f \in p\mathcal{E}^*$. U_N^q is called the *q-potential operator of* N and, as usual, we write $U_N = U_N^0$. In consulting the literature one must be careful: U_N^q is often defined with $f \circ X_t$ replaced by $f(X_{t-})$. Of course, this requires the existence in E of X_{t-}. Under our hypotheses,

(8.4) seems the more natural definition. If $N(\omega, \cdot)$ is a.s. diffuse, then the two definitions agree. The following analog of the resolvent equation is easily derived for $0 \le r < q$ and $f \in p\mathcal{E}^*$

$$(8.5)\quad U_N^r f = U_N^q f + (q-r)U^q U_N^r f = U_N^q f + (q-r)U^r U_N^q f\,,$$

but one must be careful not to subtract since we make no finiteness assumption on $U_N^q f$. See [**S**, 36.16]. Finally one defines $u_N^q = U_N^q 1$ and u_N^q is called the *q-potential of N*. Under appropriate finiteness assumptions u_N^q or U_N^q determine N or N^o (resp. N^p) where N^o (resp. N^p) is the dual optional (resp. predictable) projection of N. See (36.3) and (36.17) in [S], for example.

A *raw additive functional* (RAF) is an increasing process $A = (A_t)_{t \ge 0}$ which satisfies $A_{t+s} = A_t + A_s \circ \theta_t$ identically in $s, t \ge 0$ and $\omega \in \Omega$. Recalling that an increasing process is finite on $[0, \zeta[$ and constant on $[\zeta, \infty[$, it is immediate that $N(dt) := dA_t$ is a HRM. (If $t \ge \zeta(\omega)$ then $\zeta(\theta_t \omega) = 0$ and so $A_s(\theta_t \omega) = 0$ for all $s \ge 0$.) An *additive functional* (AF) is a RAF, A, such that $A_t \in \mathcal{F}_t$ for all $t \ge 0$. If A is a RAF and $N(dt) := dA_t$, then we write U_A^q for U_N^q, etc. If N is a HRM, then $A_t := N(]0, t])$ is a RAF provided it is finite for $t < \zeta$.

Before proceeding we record some elementary facts about s-finite measures. Recall that a measure is s-finite provided it is a countable sum of finite measures.

(8.6) **Lemma.** *Let $(M, \mathcal{M})$ be an arbitrary measurable space. Let $\mathcal{S}$ be the class of s-finite measures on $(M, \mathcal{M})$.*

- (i) *$\nu \in \mathcal{S}$ if and only if there exists a finite measure μ with $\nu = f\mu$ and $f \in p\mathcal{M}$. If $\nu \in \mathcal{S}$ and μ is a finite measure with $\nu << \mu$, then $\nu = f\mu$ with $f \in p\mathcal{M}$.*
- (ii) *Let (ν_n) be a countable subset of $\mathcal{S}$. Then there exists a unique $\lambda \in \mathcal{S}$ with $\nu_n \le \lambda$ for all n and such that if $\gamma \in \mathcal{S}$ and $\nu_n \le \gamma$ for all n, then $\lambda \le \gamma$. We write $\lambda = \sup \nu_n$.*

(iii) *Let* (ν_n) *be an increasing sequence in* $\mathcal{S}$ *and let* $\nu := \uparrow \lim \nu_n$. *Then* $\nu \in \mathcal{S}$ *and* $\nu = \sup \nu_n$.

Proof. (i) Obviously any measure of the form $f\mu$ with μ finite and $f \in p\mathcal{M}$ is in $\mathcal{S}$. Conversely given $\nu \in \mathcal{S}$, then $\nu = \sum \nu_n$ where each ν_n is finite. We may suppose $\nu_n(M) > 0$ for each n. Let $\mu = \sum 2^{-n}(\nu_n(M))^{-1}\nu_n$. Then μ is a finite measure and $\nu << \mu$. Now let μ be any finite measure with $\nu << \mu$. Then $\nu_n << \mu$ for each n and so $\nu_n = f_n\mu$ with $f_n \in p\mathcal{M}$. If $f = \sum f_n$, then $\nu = f\mu$ proving (i).

(ii) Clearly there exists a finite measure μ such that $\nu_n = f_n\mu$ with $f_n \in p\mathcal{M}$ for each n. Let $f := \sup f_n$ and $\lambda := f\mu$. Then $\lambda \in \mathcal{S}$ and $\nu_n \leq \lambda$ for each n. Let $\gamma \in \mathcal{S}$ and $\gamma \geq \nu_n$ for each n. Then there exists a finite measure $\bar{\mu}$ with $\gamma = \bar{h}\,\bar{\mu}$. Set $\rho = \mu + \bar{\mu}$ and write $\mu = \varphi\rho$ and $\bar{\mu} = \psi\rho$. Define $g_n = f_n\varphi$, $g = f\varphi$, and $h = \bar{h}\psi$ so that $\nu_n = g_n\rho$, $\lambda = g\rho$, and $\gamma = h\rho$. Since $\nu_n \leq \gamma$, $g_n \leq h$ a.e. ρ and so $g \leq h$ a.e. ρ. Thus $\lambda \leq \gamma$, proving (ii).

(iii) Again write $\nu_n = f_n\mu$ with μ finite measure and $f := \sup f_n$. Since (ν_n) is increasing, $f_n \uparrow f$ a.e. μ and $\nu = f\mu$. Hence $\nu \in \mathcal{S}$. Let $\lambda = \sup \nu_n$. Then $\lambda \leq \nu$. But $\nu_n \leq \lambda$ and $\nu_n \uparrow \nu$ imply that $\nu \leq \lambda$. ∎

Remark. If $\nu \in \mathcal{S}$, and $\nu \neq 0$, then $\nu = f\mu$ with μ a probability. Then $\nu_1 := f\,1_{\{f<\infty\}}\mu$ is σ-finite and $\nu_2 := 1_{\{f=\infty\}}\mu$ is a finite measure such that $\nu = \nu_1 + \infty \cdot \nu_2$ and ν_1 and ν_2 are carried by disjoint sets.

Fix a HRM, N and $q \geq 0$. If $\xi \in \text{Exc}^q$, define $\varphi(t) = P^\xi \int_{]0,t]} e^{-qs}N(dt)$. Then using (8.3) one has

$$(8.7)\quad \varphi(t+s) = \varphi(t) + e^{-qt}P^{\xi P_t}\int_{]0,s]} e^{-qs}N(ds) \leq \varphi(t) + \varphi(s)$$

because $\xi \in \text{Exc}^q$. The following lemma is now proved exactly as (75.5) in [**S**]. See [**FG88**] for a slightly different proof and also (10.39).

(8.8) **Lemma.** *Either φ is identically infinite or φ is a finite, continuous, increasing, concave function on $[0,\infty[$ with $\varphi(0) = 0$. In particular, $\varphi(t)/t$ increases to a limit $L \in [0,\infty]$ as t decreases to zero. One also has $L = \uparrow \lim_{p\to\infty} p\xi(u_N^{p+q})$.*

If $f \in p\mathcal{E}$ we may apply (8.8) to $f * N(dt) := f \circ X_t N(dt)$, which is a HRM whenever N is, to define

$$(8.9) \qquad {}^q\nu_N^\xi(f) := \uparrow \lim_{t\downarrow 0} t^{-1} P^\xi \int_{]0,t]} e^{-qs} f \circ X_s N(dt)$$

$$= \uparrow \lim_{p\to\infty} p\langle \xi, U_N^{p+q} f\rangle \,.$$

It is clear that ${}^q\nu_N^\xi$ is a measure on E that is additive and positive homogeneous in both ξ and N. We shall call ${}^q\nu_N^\xi$ the *Revuz measure of N relative to* $\xi \in \mathrm{Exc}^q$. Of course, when $q = 0$ we drop it from our notation. We caution the reader once again that the terminology varies in the literature. Often the Revuz measure is defined with $f \circ X_s$ replaced by $f(X_{s-})$ in (8.9), while what we have defined as the Revuz measure in (8.9) is sometimes called the characteristic measure of N relative to ξ. Observe that if $N(ds) = 1_{]0\zeta[}(s)\,ds$, then $U_N^q = U^q$, and since $p\xi U^{p+q} \uparrow \xi$ as $p \to \infty$ when $\xi \in \mathrm{Exc}$, ${}^q\nu_N^\xi = \xi$ for all $q \geq 0$ in this case. More generally, suppose $\xi \in \mathrm{Exc}^r$. If $q > r$, then $\xi \in \mathrm{Exc}^q$ and

$$(8.10) \qquad {}^q\nu_N^\xi(f) = \lim_{p\to\infty} p\langle \xi, U_N^{q+p} f\rangle$$

$$= \lim_{p\to\infty} p\langle \xi, U_N^{r+p} f\rangle = {}^r\nu_N^\xi(f)\,.$$

In particular if $\xi \in \mathrm{Exc}$, then ${}^q\nu_N^\xi = \nu_N^\xi$ for all $q \geq 0$. Note that if $\xi \in \mathrm{Inv}^q$, then (8.7) becomes $\varphi(t+s) = \varphi(t) + \varphi(s)$, and consequently ${}^q\nu_N^\xi(f) = P^\xi \int_{]0,1]} e^{-qt} f \circ X_t N(dt)$ in this case. Naturally we write ${}^q\nu_A^\xi$ when A is a RAF. Let N be a HRM and $\xi \in \text{Exc}^q$. Then we say that N is *ξ-integrable* provided ${}^q\nu_N^\xi(1) < \infty$ and that N is *ξ-σ-integrable* provided ${}^q\nu_N^\xi$ is σ-finite. Using the N_k in (8.1) and the σ-finiteness of ξ it is clear that $f \to p\langle \xi, U_N^{p+q} f\rangle$ is s-finite, and consequently by (8.6), so is ${}^q\nu_N^\xi$. Therefore we may use the Fubini theorem in computations involving Revuz measures.

(8.11) **Proposition.** *Let N be a* HRM *and $\xi = \mu U^q \in \text{Pot}^q(X)$, $q \geq 0$. Then ${}^q\nu_N^\xi = \mu U_N^q$.*

Proof. Using the Fubini theorem we find

$$\varphi(t) := P^{\mu U^q} N(]0,t]) = P^\mu \int_0^\infty e^{-qs} N(]s, t+s])\, ds$$

$$= P^\mu \int N(dr) \int_{(r-t)^+}^{r} e^{-qs}\, ds\,.$$

But $t^{-1} \int_{(r-t)^+}^{r} e^{-qs}\, ds \to e^{-qr}$ as $t \to 0$ when $r > 0$, and is dominated by

$$t^{-1} e^{-q(r-t)^+}[r - (r-t)^+] \leq e^{-q(r-t)^+} \leq e^{qt} e^{-qr} \leq e^q e^{-qr}$$

provided $t \leq 1$. Consequently, by the dominated convergence theorem $t^{-1}\varphi(t)$ approaches $\mu(u_N^q)$ whenever $\mu(u_N^q) < \infty$. Suppose $\mu(u_N^q) = \infty$. Observe that

$$t^{-1} \int_{(r-t)^+}^{r} e^{-qs}\, ds \geq t^{-1} e^{-qr}[r - (r-t)^+]\,.$$

But $t^{-1}[r-(r-t)^+] \uparrow 1$ as $t \downarrow 0$ if $r > 0$, and so $\liminf_{t\to 0} t^{-1}\varphi(t) \geq \mu(u_N^q) = \infty$. Therefore $t^{-1}\varphi(t) \uparrow \mu(u_N^q)$ as $t \downarrow 0$ in all cases. Applying this to $f * N$ for $f \in p\mathcal{E}$ gives (8.11). ∎

(8.12) **Remark.** Actually (8.8) holds for any s-finite measure ξ which satisfies $\xi P_t \leq \xi$ for all t. Since the only additional thing used in the proof of (8.11) is the Fubini theorem one finds

$$\uparrow \lim_{t\downarrow 0} t^{-1} P^{\mu U^q} \int_{]0,t]} f \circ X_s N(ds) = \mu U_N^q(f)$$

for any s-finite μ.

The next result is taken from **[Me73]**.

(8.13) **Proposition.** *Let* N *be a* HRM *and* $\xi \in \mathrm{Dis}^q$. *Then* ${}^q\nu_N^\xi(f) = L^q(\xi, U_N^q f)$. *Of course,* $\mathrm{Dis}^q = \mathrm{Exc}^q$ *if* $q > 0$.

Proof. Since the proof is the same for all $q \geq 0$ we shall write it only for $q = 0$. Because $\xi \in \mathrm{Dis}$ there exist potentials $\mu_n U \uparrow \xi$. Given $f \in p\mathcal{E}$, $U_N f$ is excessive, and so by (2.14) there exist potentials $U f_k \uparrow U_N f$ a.e. ξ and a.e. μ_n for each n. It is immediate from the definition (8.9) that $\nu_N^{\mu_n U} \uparrow \nu_N^\xi$. Therefore using (8.11)

$$
\begin{aligned}
\nu_N^\xi(f) &= \uparrow \lim_n \mu_n U_N(f) = \uparrow \lim_n \uparrow \lim_k \mu_n(U f_k) \\
&= \uparrow \lim_k L(\xi, U f_k) = L(\xi, U_N f) \qquad \blacksquare
\end{aligned}
$$

If N is a HRM, then a set $B \in \mathcal{E}^*$ *carries* N provided $1_B * N = N$. If N is carried by B, then so is ${}^q\nu_N^\xi$ for $\xi \in \mathrm{Exc}^q$, $q \geq 0$. Observe that if N is diffuse and carried by B, then almost surely N is carried by $]T_B, \infty[$ (that is, $1_{]T_B,\infty[} * N = N$) and it follows that $P_B^q U_N^q = U_N^q$ provided $B \in \mathcal{E}^e$.

(8.14) **Proposition.** *Let* N *be a* HRM *and* T *be an exact terminal time such that* $P_T^q U_N^q = U_N^q$. *Then for each* $\xi \in \mathrm{Exc}^q$, ${}^q\nu_N^\xi = {}^q\nu_N^\eta$ *where* $\eta = R_T^q \xi$.

Proof. It suffices to consider the case $q = 0$. First, if $\xi \in \mathrm{Dis}$, then using (8.13) and (4.12),

$$
\begin{aligned}
\nu_N^\xi(f) &= L(\xi, U_N f) = L(\xi, P_T U_N f) \\
&= L(R_T \xi, U_N f) = \nu_N^{R_T \xi}(f).
\end{aligned}
$$

If $\xi \in \mathrm{Con}$, then $\xi \in \mathrm{Dis}^q$ for $q > 0$. Therefore by what we have just proved and (8.10), $\nu_N^\xi = {}^q\nu_N^\xi = {}^q\nu_N^\eta$ where $\eta = R_T^q \xi$. Hence

$$
\nu_N^\xi(f) = \uparrow \lim_{p \to \infty} p \langle R_T^q \xi, U_N^{p+q} f \rangle .
$$

But $R_T^q \xi \uparrow R_T \xi$ by (4.7) and $U_N^{p+q} f \uparrow U_N^p f$ as $q \downarrow 0$. Consequently letting $q \downarrow 0$ in the last display we obtain

$$\nu_N^\xi(f) = \uparrow \lim_{p\to\infty} p\langle R_T\xi, U_N^p f\rangle = \nu_N^{R_T\xi}(f).$$

Proposition (8.14) now follows because both $\xi \to \nu_N^\xi$ and $\xi \to R_T\xi$ are additive. ∎

As we have seen ν_N^ξ is s-finite in general. However in an important special case it is actually σ-finite. Note that the possible mass of N at ζ does not influence ν_N^ξ; that is, N and $1_{]0,\zeta[} * N$ have the same Revuz measure. The following result is due to Revuz **[Re70]**. Since the argument given in [**S**] to prove (75.9) carries over without change to prove it, we shall not repeat it here.

(8.15) **Proposition.** *Let A be an additive functional with uniformly bounded jumps, that is,* $\sup_\omega \sup_{t<\zeta(\omega)} [A_t(\omega) - A_{t-}(\omega)] < \infty$. *Then for each $\xi \in \mathrm{Exc}$, ν_A^ξ is σ-finite.*

(8.16) **Remark.** We should point out that the argument in [**S**, (75.9] does not prove what is claimed there. In the remainder of this remark the notation is that of [**S**, §75]. What is established is the following. If A is an AF ($A_t < \infty$ if $t < \zeta$) not charging ζ and having uniformly bounded jumps, then $E = \cup E_n$ where, defining $A_n := 1_{E_n} * A$, each A_n has a finite Revuz measure relative ξ. Consequently ν_A^ξ is σ-finite. This does not require A to be predictable or u_A^1 bounded or even finite. However, if A is predictable, then in [**S**, (75.6)] with $\kappa = d A_n$ one may replace $1_{E_n} \circ X_t$ by ${}^p(1_{E_n} \circ X_t)$ to obtain the statement in (75.9-ii) *without* assuming that u_A^1 is bounded. Note Sharpe's ν_κ^ξ is defined on $\mathcal{H}^g$ and is quite different from our definition of ν_κ^ξ.

We now turn to the task of expressing the Revuz measure ν_N^ξ in terms of the Kuznetsov measure Q_ξ. Therefore in the remainder of this section we assume that (6.2) is in force.

Recall that, in particular, this means that X is realized on the space Ω defined below (6.5). Our first task is to extend a HRM, N, of X defined on Ω to W. The technique for this goes back to Mitro [**Mi79**]. See also [**GS84**]. Recall that we regard $N(\omega, \cdot)$ as a measure on $\mathbb{R}$ that is carried by $]0, \zeta(\omega)]$. Thus for each $t \in \mathbb{R}$

$$(8.17) \qquad \bar{N}_t(w, B) := 1_{\{\alpha(w)<t\}} N(\theta_t w, B - t)$$

is well defined on $\mathcal{B} := \mathcal{B}(\mathbb{R})$ for each $w \in W$ and is carried by $]t, \beta(w)]$. Since θ_t as a map from $\{\alpha < t\}$ to Ω is in $\mathcal{G}^*_{\geq t}|\mathcal{F}^*$ where $\mathcal{G}^0_{\geq t} = \sigma\{Y_s : s \geq t\}$ and $\mathcal{G}^*_{\geq t}$ is its universal completion, $\bar{N}_t$ is a kernel from $(W, \mathcal{G}^*_{\geq t})$ to $(\mathbb{R}, \mathcal{B})$ for each t. Clearly $\bar{N}_t = \sum \bar{N}^k_t$ where $\bar{N}^k_t$ is the *bounded* kernel defined from N_k in the definition (8.1) by means of (8.17). We now fix $w \in W$, but suppress it in our notation. If $\alpha < s < t$ and $B \subset]t, \infty[$, then

$$N(\theta_s, B - s) = N(\theta_{t-s}\theta_s, B - s - (t - s)) = N(\theta_t, B - t),$$

where the first equality comes from the homogeneity of N on Ω. Consequently for $B \in \mathcal{B}$ and $\alpha < s < t$, letting $B_t = B \cap]t, \infty[$, we have

$$N(\theta_s, B - s) \geq N(\theta_s, B_t - s) = N(\theta_t, B - t)$$

with equality if $B \subset]t, \infty[$. Therefore

$$(8.18) \qquad N^*(w, B) := \sup_{r \in \mathbb{Q}, r > \alpha(w)} N(\theta_r w, B - r) = \uparrow \lim_{t \downarrow \alpha(w)} N(\theta_t w, B - t)$$

defines for each $w \in W$ an s-finite measure on $(\mathbb{R}, \mathcal{B})$ because of (8.6-ii) which is carried by $]\alpha(w), \beta(w)]$. Clearly N^* is a kernel from $(W, \mathcal{G}^*)$ to $(\mathbb{R}, \mathcal{B})$. Moreover if $B \subset]t, \infty[$, then $1_{\{\alpha<t\}} N^*(B) = 1_{\{\alpha<t\}} N(\theta_t, B - t)$ is $\mathcal{G}^*_{\geq t}$ measurable.

Also N^* is the supremum of the countable family of kernels $\bar{N}_r^k := 1_{\{\alpha<r\}}N_k(\theta_r, B-r)$ as r ranges over $\mathbb{Q}$ and $k \geq 1$. Now an argument similar to the proof of (8.6-ii) and using Doob's lemma, [**S**, (A3.2)], shows that N^* is a countable sum of bounded kernels from $(W, \mathcal{G}^*)$ to $(\mathbb{R}, \mathcal{B})$. Finally observe that for each $t \in \mathbb{R}$ and $w \in W$

$$\begin{aligned} N^*(\sigma_t w, B) &= \uparrow \lim_{s \downarrow \alpha(w)-t} N(\theta_s \sigma_t w, B - s) \\ &= \uparrow \lim_{u \downarrow \alpha(w)} N(\theta_u w, B + t - u) = N^*(w, B+t), \end{aligned}$$

for all $B \in \mathcal{B}$.

This leads to the following definition.

(8.19) **Definition.** A *homogeneous random measure of* $Y, K = (K(w, dt))$, is a kernel from $(W, \mathcal{G}^*)$ to $(\mathbb{R}, \mathcal{B})$ that is a countable sum of bounded kernels and satisfies: (i) for each $w \in W$, $K(w, \cdot)$ is carried by $[\alpha(w), \beta(w)]$; and (ii) $K(\sigma_t w, B) = K(w, B+t)$ for each $t \in \mathbb{R}$, $w \in W$, and $B \in \mathcal{B}$.

Thus if N is a HRM of X, then N^* defined in (8.18) is a HRM of Y which is carried by $]\alpha, \beta]$ and has the additional property that

$$N^*(w, B) = N(\theta_t w, B - t) \tag{8.20}$$

when $B \subset]t, \infty[$ and $\alpha(w) < t$. In particular, $1_{\{\alpha<t\}}N^*(B) \in \mathcal{G}^*_{\geq t}$ when $B \subset]t, \infty[$. Of course, (8.18) implies for $\varphi \in p\mathcal{B}$

$$\int \varphi(s) N^*(w, ds) = \uparrow \lim_{t \downarrow \alpha(w)} \int \varphi(t+s) N(\theta_t w, ds).$$

Moreover, using this, it is readily checked that the extension of $f * N := f \circ X_t N(dt)$ by (8.18) is $f \circ Y_t N^*(dt)$.

We now drop the "$*$" in our notation and use the same symbol to denote the extension of a HRM of X to the HRM of Y defined in (8.18). Observe that if S is a stationary time as defined in (6.21) and (7.8), then $K(dt) := 1_{[\alpha,\beta]}(S)\epsilon_S(dt)$ defines a HRM of Y.

(8.21) **Theorem.** *Let N be a* HRM *of X and $\xi \in$ Exc. If $\varphi \in p\mathcal{B}$ with $\int \varphi(t)\,dt = 1$, then for $f \in p\mathcal{E}^*$*

$$\nu_N^\xi(f) = Q_\xi \int \varphi(t) f \circ Y_t N(dt).$$

Proof. Suppose first that $\xi \in$ Dis. Fix $f \in p\mathcal{E}^*$. Then according to (8.13), $\nu_N^\xi(f) = L(\xi, U_N f)$. Define $H := \int_0^\infty f \circ X_t N(dt)$. Then $H \in \mathcal{F}^*$ and $P^x(H) = U_N f(x)$. It follows that H is excessive, and from (8.18) that H^* defined in (7.1) is given by $H^* = \int f \circ Y_s N(ds)$. Consequently by (7.2), $\nu_N^\xi(f) = Q_\xi(H^*; 0 < S < 1)$ where S is a stationary time with $\alpha < S < \beta$ a.s. Q_ξ. Now let $\varphi \in p\mathcal{B}$ with $\int \varphi(t)\,dt = 1$. Define $G := \int \varphi(t) f \circ Y_t N(dt) \in \mathcal{G}^*$. Then

$$\begin{aligned} \bar{G} := \int G \circ \sigma_t\,dt &= \int dt \int \varphi(s) f \circ Y_{t+s} N(ds + t) \\ &= \int dt \int \varphi(s-t) f \circ Y_s N(ds) = \int f \circ Y_s N(ds) = H^*. \end{aligned}$$

Therefore, using (6.26) and (6.27), we have

$$\nu_N^\xi(f) = Q_\xi(H^*; 0 < S < 1) = Q_\xi \int \varphi(t) f \circ Y_t N(dt),$$

proving (8.21) when $\xi \in$ Dis. Next suppose $\xi \in$ Con $\subset$ Inv. As remarked below (8.10) $\nu_N^\xi(f) = P^\xi \int_{]0,1]} f \circ X_t N(dt)$ in this case. Let $F := \int_{]0,1]} f \circ X_t N(dt) \in p\mathcal{F}^*$. It follows from (8.20) that $F \circ \theta_0 = \int_{]0,1]} f \circ Y_t N(dt)$ a.s. Q_ξ since $\alpha = -\infty$ a.s. Q_ξ.

Therefore

$$(8.22)\quad Q_\xi \int_{]0,1]} f\circ Y_t N(dt) = Q_\xi(F\circ\theta_0)$$
$$= Q_\xi P^{Y(0)}(F) = P^\xi(F) = \nu_N^\xi(f)\,.$$

In order to complete the proof of (8.21) we need a lemma which we now state but whose proof is postponed to the end of this section.

(8.23) **Lemma.** *Let $(M,\mathcal{M})$ be a measurable space and let μ be an s-finite measure on $(\mathbb{R}\times M, \mathcal{B}\otimes\mathcal{M})$ that is translation invariant along $\mathbb{R}$; that is, if $H\in p(\mathcal{B}\otimes\mathcal{M})$, then $\mu(H_s) = \mu(H)$ for $s\in\mathbb{R}$ where $H_s(t,x) := H(s+t,x)$. Then there exists a unique s-finite measure, ν, on $(M,\mathcal{M})$ such that $\mu = \ell\times\nu$ where ℓ is Lebesgue measure on $\mathbb{R}$; that is, for $h\in p(\mathcal{B}\otimes\mathcal{M})$*

$$(8.24)\quad \mu(h) = \int dt \int h(t,x)\nu(dx) = \int \nu(dx)\int h(t,x)\,dt\,.$$

We now return to the proof of (8.21). This part of the argument does not require $\xi\in\text{Con}$ and is valid for any HRM, N, of Y. For $h\in p(\mathcal{B}\otimes\mathcal{E}^*)$ define $\mu(h) := Q_\xi\int h(t,Y_t)N(dt)$. Then μ is an s-finite measure on $\mathcal{B}\otimes\mathcal{E}^*$. Using the homogeneity of N and the stationarity of Q_ξ one sees that μ is translation invariant along $\mathbb{R}$. Let ν be the s-finite measure on $(E,\mathcal{E}^*)$ whose existence is asserted in (8.23). Taking $h(t,x) = 1_{]0,1]}(t)f(x)$ in (8.24), one obtains $\nu(f) = Q_\xi \int_{]0,1]} f\circ Y_t N(dt)$. Combining this with (8.22) and (8.24) completes the proof of (8.21). ∎

The argument in the last paragraph of the proof of (8.21) establishes the following result.

(8.25) **Proposition.** *Let* N *be a* HRM *of* Y *and* $\xi \in \text{Exc}$. *Then* $\nu_N^\xi(f) := Q_\xi \int_{]0,1]} f \circ Y_t N(dt)$ *defines an* s*-finite measure on* $\mathcal{E}^*$ *such that*

$$Q_\xi \int F(t, Y_t) N(dt) = \int_{-\infty}^{\infty} dt \int F(t,x) \nu_N^\xi(dx)$$

for all $F \in p(\mathcal{B} \otimes \mathcal{E}^*)$.

Remark. We shall call ν_N^ξ the *Revuz measure of* N *relative to* ξ. In view of (8.21) this is consistent with our previous definition (8.9) when N is a HRM of X.

It remains to prove (8.23). This is proved in [**DM**, XVIII-24]. Nevertheless we shall give a proof. See also [**G87**]. Suppose first that μ is an s-finite translation invariant measure on $(\mathcal{B}, \mathbb{R})$; that is, the special case in which M reduces to a single point. If μ is σ-finite it is a standard fact that $\mu = c \cdot \ell$ where $0 \le c < \infty$. This remains true when μ is only s-finite except that $c = \infty$ is possible. To see this let $\varphi, \psi \in p\mathcal{B}$. Then

$$\begin{aligned} \mu(\varphi)\ell(\psi) &= \int \psi(t) \int \varphi(s-t)\mu(ds)\, dt \\ &= \int \mu(ds) \int \varphi(t)\psi(s-t)\, dt = \mu(\psi)\ell(\varphi), \end{aligned}$$

where the first equality uses the translation invariance of μ and second the fact that ℓ is translation invariant and is also invariant under reflection in the origin. Choose $\psi \in p\mathcal{B}$ with $\ell(\psi) = 1$, to obtain $\mu = \mu(\psi)\ell$.

We now consider the general case in (8.23). If $f \in p\mathcal{M}$, then $\varphi \to \mu(\varphi \otimes f)$ defines an s-finite translation measure on $\mathbb{R}$ where $\varphi \otimes f(t,x) := \varphi(t) f(x)$. Hence there exists a constant $\nu(f)$ such that $\mu(\varphi \otimes f) = \nu(f)\ell(\varphi)$. Choosing $\varphi \in p\mathcal{B}$ with $\ell(\varphi) = 1$, it is clear that ν is an s-finite measure on $(M, \mathcal{M})$. Let μ' be the s-finite measure on $\mathbb{R} \times M$ defined by (8.24). (The second equality in (8.24) is just the Fubini theorem.)

Given $\varphi \in p\mathcal{B}$ with $\ell(\varphi) = 1$ and $F \in p(\mathcal{B} \otimes \mathcal{M})$ define $\varphi * F(t,x) := \int \varphi(t-s)F(s,x)\,ds = \int \varphi(s)F(t-s,x)\,ds$. Then

$$\mu(\varphi * F) = \int ds\,\varphi(s) \iint F(t-s,x)\mu(dt,dx) = \mu(F).$$

Consequently using $\mu(\psi \otimes f) = \nu(f)\ell(\psi)$ for the second equality

$$\begin{aligned}\mu(F) &= \int ds \iint \varphi(t-s)F(s,x)\mu(dt,dx) \\ &= \int ds \iint \varphi(t-s)F(s,x)\,dt\,\nu(dx) = \mu'(F),\end{aligned}$$

completing the proof of (8.23).

(8.26) **Remark.** Even if μ is a σ-finite measure on $\mathbb{R} \times M$ that is translation invariant along $\mathbb{R}$, then all that is a priori clear is that $\mu(\varphi \otimes f) = \ell(\varphi)\nu(f)$ where ν is s-finite. This is not enough to allow one to use the usual uniqueness theorem for product measures to conclude $\mu = \ell \times \nu$. However, this is true by (8.23). If $F > 0$ with $\mu(F) < \infty$, then $f(x) := \int F(t,x)\,dt > 0$ and $\nu(f) < \infty$. Therefore ν is, in fact, σ-finite. However, I do not see how to prove this without using something like (8.23). See example (iii) on page 32 of **[G87]** in this connection.

9. Flows and Palm Measures

Portions of the material developed in sections 6 and 8 make sense in the more general context of a measure preserving flow without any Markovian structure. We shall develop some of these ideas in this section and apply them to Markov processes in §10. We shall use the notation of our main application—the flow associated with a Kuznetsov measure.

(9.1) Definition. A *flow* $(W, \mathcal{G}^0, Q, (\sigma_t)_{t\in\mathbb{R}})$ is a σ-finite measure space $(W, \mathcal{G}^0, Q)$ equipped with a family $(\sigma_t)_{t\in\mathbb{R}}$ of measurable mappings from W to W such that: (i) $\sigma_{t+s} = \sigma_t\sigma_s$ and σ_0 is the identity; (ii) $(t,w) \to \sigma_t(w)$ as a map from $\mathbb{R}\times W$ to W is $\mathcal{B}\otimes\mathcal{G}^0/\mathcal{G}^0$ measurable; (iii) $\sigma_t(Q) = Q$ for all $t \in \mathbb{R}$.

It is immediate that each σ_t is a bijection of W and that $\sigma_t^{-1} = \sigma_{-t}$, $t \in \mathbb{R}$. It follows from (9.1ii) that if $f \in \mathcal{G}^* := (\mathcal{G}^0)^*$, then $(t,w) \to f(\sigma_t w)$ is $(\mathcal{B}\otimes\mathcal{G}^0)^*$ measurable. More generally, if $F \in (\mathcal{B}\otimes\mathcal{G}^0)^*$, then $(t,w) \to F(t,\sigma_t w)$ is $(\mathcal{B}\otimes\mathcal{G}^0)^*$ measurable. In particular if $\mathcal{G}^0$ contains all singletons $\{w\}$, then $t \to F(t,\sigma_t w)$ is $\mathcal{B}^*$ measurable for each $w \in W$. In any case there is more than enough joint measurability to use the Fubini theorem with impunity. If the underlying measure space $(W, \mathcal{G}^0, Q)$ is understood, a family $(\sigma_t)_{t\in\mathbb{R}}$ satisfying conditions (i), (ii), and (iii) of (9.1) is called a flow on $(W, \mathcal{G}^0, Q)$. In the remainder of this section $(W, \mathcal{G}^0, Q, (\sigma_t)_{t\in\mathbb{R}})$ will be a fixed flow.

Example. Let X be a right process satisfying (6.2) and W, $\mathcal{G}^0$, and σ_t be defined as in the third paragraph of §6. Given $\xi \in \text{Exc}$, then $(W, \mathcal{G}^0, Q_\xi, (\sigma_t)_{t\in\mathbb{R}})$ is a flow. We shall call it the *Kuznetsov flow* of ξ.

Remark. A considerable part of what follows in this section is valid if Q is only assumed to be s-finite rather than σ-

finite. Since our main application is to Kuznetsov flows we shall make the blanket hypothesis that Q is σ-finite and leave it to the interested to sort out those results which are valid for an s-finite Q.

Let $\mathcal{I}$ be the σ-algebra of invariant sets; that is, $F \in \mathcal{I}$ provided $F \in \mathcal{G}^*$ and $F(\sigma_t w) = F(w)$ identically in t and w. Given $F \in p\mathcal{G}^*$ define $\bar{F} := \int^* F \circ \sigma_t \, dt$. Then by (B.11), $\bar{F} \in \mathcal{I}$. The following proposition follows immediately from Fubini's theorem exactly as (6.27).

(9.2) Proposition. *Let* $F, G \in p\mathcal{G}^*$ *and* $H \in p\mathcal{I}$. *Then* $Q(F\bar{G}H) = Q(\bar{F}GH)$.

We next define the analog for flows of a HRM.

(9.3) Definition. A *stationary measure* (SM) is a kernel, N, from $(W, \mathcal{G}^*)$ to $(\mathbb{R}, \mathcal{B})$ that is a countable sum of bounded kernels N_k of the same type and such that

$$N(\sigma_t w, B) = N(w, B + t), \tag{9.4}$$

identically in $t \in \mathbb{R}$, $w \in W$, and $B \in \mathcal{B}$.

We emphasize that the bounded kernels N_k in (9.3) are *not* assumed to have the property (9.4). Just as in the definition (8.1) of a RM, the condition $N = \sum N_k$ where each N_k is bounded is equivalent to the apparently weaker condition that each N_k be finite; that is, $N_k(w, \mathbb{R}) < \infty$. If $(W, \mathcal{G}^0, Q_\xi, (\sigma_t)_{t \in \mathbb{R}})$ is a Kuznetsov flow then any HRM as defined in (8.19) is a SM. We shall say that two SMs N_1 and N_2 are *equivalent* provided $N_1(w, \cdot) = N_2(w, \cdot)$ as measures on $(\mathbb{R}, \mathcal{B})$, a.s. Q. (If more than one measure is under consideration we shall write Q-equivalent for clarity.)

(9.5) Remark. There is a "perfection" theorem for stationary measures that satisfy an additional finiteness condition. See (9.28), (9.34), and (9.35) for variations on this theme.

If N is a SM and $\varphi \in p\mathcal{B}$, define $N(\varphi) := \int \varphi(t) N(dt)$. Then $N(\varphi) \in p\mathcal{G}^*$ and $\overline{N(\varphi)} := \int N(\varphi) \circ \sigma_t \, dt = \ell(\varphi) N(\mathbb{R})$

where, as usual, ℓ is Lebesgue measure on $\mathbb{R}$. Consequently given SMs N_1 and N_2, $H \in p\mathcal{I}$, and $\varphi_1, \varphi_2 \in p\mathcal{B}$, (9.2) implies

$$(9.6)\quad Q[N_1(\varphi_1)H\, N_2(\mathbb{R})]\ell(\varphi_2) = \ell(\varphi_1)Q[N_1(\mathbb{R})H\, N_2(\varphi_2)].$$

(9.7) Definition. A *stationary time* (ST) is a map $S\colon W \to [-\infty, \infty]$ which is $\mathcal{G}^*$ measurable and such that $S(\sigma_t w) = S(w) - t$ identically.

Obviously $N(dt) := 1_{\mathbb{R}}(S)\epsilon_S(dt)$ is a SM whenever S is a ST. If S is a ST and $\varphi \in p\mathcal{B}$, than as in §6, $\overline{\varphi(S)} = \ell(\varphi)\, 1_{\mathbb{R}}(S)$. Thus (9.6) becomes

$$(9.8)\qquad Q[\varphi(S)H;\, T \in \mathbb{R}]\ell(\psi) = \ell(\varphi)Q[H\psi(T);\, S \in \mathbb{R}]$$

for any two STs S and T, $H \in p\mathcal{I}$, and $\varphi, \psi \in p\mathcal{B}$.

(9.9) Proposition. *Let N be a SM. Then there exists a unique s-finite measure P_N on $(W, \mathcal{G}^0)$ such that for $F \in p(\mathcal{B} \otimes \mathcal{G}^0)^*$*

$$(9.10)\qquad \int Q(dw) \int F(t, \sigma_t w) N(w,\, dt) = \int dt \int F(t, w) P_N(dw).$$

If $f \in p\mathcal{G}^*$ and $\varphi \in p\mathcal{B}$, then

$$(9.11)\qquad \ell(\varphi)P_N(f) = \int Q(dw) \int \varphi(t) f(\sigma_t w) N(w, dt).$$

P_N is called the Palm measure of N (relative to Q).

Remark. In the future we shall omit the w's in formulas such as (9.10) or (9.11). For example, (9.10) will be written

$$Q \int F(t, \sigma_t) N(dt) = \int dt\, P_N(F(t, \cdot)).$$

Proof. Letting $F(t,w)=\varphi(t)f(w)$ in (9.10) gives (9.11). For the existence of P_N define η on $\mathcal{B}\otimes\mathcal{G}^0$ by the left hand side of (9.10). Then η is an s-finite measure. If $F\in p(\mathcal{B}\otimes\mathcal{G}^0)$, let $F_s(t,w)=F(t+s,w)$. Then

$$\begin{aligned}\eta(F_s)&=Q\int F(s+t,\sigma_t)N(dt)\\&=Q\int F(t,\sigma_{t-s})N(\sigma_{-s},dt)=\eta(F),\end{aligned}$$

where the second equality uses the stationarity of N and the last equality the invariance of Q under σ_{-s}. Consequently by (8.23) there exists a unique s-finite measure, P_N, such that $\eta(F)$ is given by the right hand side of (9.10) first for $F\in p(\mathcal{B}\otimes\mathcal{G}^0)$ and, hence, for $F\in p(\mathcal{B}\otimes\mathcal{G}^0)^*$. ∎

(9.12) Remarks. If we put $\varphi=1_{\{t\}}$ and $f=1$ in (9.11) we see that $N(\{t\})=0$ a.s. Q for each t. Now let $\varphi=1_{]0,1]}$ in (9.11) to obtain for $f\in p\mathcal{G}^*$

$$P_N(f)=Q\int_0^1 f\circ\sigma_t\,N(dt). \tag{9.13}$$

A SM, N, is *σ-integrable* provided there exists $H\in p(\mathcal{B}\otimes\mathcal{G}^0)$ with $H>0$ and $Q\int H(t,\cdot)N(dt)<\infty$. Of course, no generality is gained by supposing only that $H\in p(\mathcal{B}\otimes\mathcal{G}^0)^*$. Taking $F(t,w)=H(t,\sigma_{-t}w)$ we see that the measure η defined by the left hand side of (9.10) is σ-finite when N is σ-integrable. It now follows from (8.26) that P_N is σ-finite if and only if N is σ-integrable. Suppose only that there exists $H\in\mathcal{B}\otimes\mathcal{G}^*$ with $H>0$ and $h=\int H(t,\cdot)N(dt)<\infty$ a.s. Q. Let $W_N:=\{w: N(w,\cdot)\neq 0\}\in\mathcal{G}^*$. Define

$$\begin{aligned}\gamma(t,w)&=H(t,w)/h(w) && w\in W_N \text{ and } h(w)<\infty,\\&=1 && w\notin W_N \text{ or } h(w)=\infty.\end{aligned} \tag{9.14}$$

Then $\gamma > 0$ and $\int \gamma(t,\cdot)N(dt) = 1_{W_N}$ a.s. Q. Let $g \in \mathcal{G}^0$ with $g > 0$ and $Q(g) < \infty$. Now defining $G(t,w) := g(w)\gamma(t,w) > 0$, then $G \in p(\mathcal{B} \otimes \mathcal{G}^*)$ and

$$Q \int G(t,\cdot)N(dt) \le Q(g) < \infty .$$

Thus this apparently weaker assumption implies that N is σ-integrable. It is now easily checked that if $\mathbb{R} = \cup B_k$ where $B_k \in \mathcal{B}$ and $N(\cdot, B_k) < \infty$ a.s. Q for each $k \ge 1$, then N is σ-integrable. Suppose N is σ-integrable. Given $f \in p\mathcal{G}^*$, let $F(t,w) = f(\sigma_{-t}w)\gamma(t,\sigma_{-t}w)$ in (9.10) where γ is defined in (9.14) to obtain

$$Q(f;\, W_N) = \int dt\, P_N[f(\sigma_{-t})\gamma(t,\sigma_{-t})] . \tag{9.15}$$

Therefore P_N and γ determine the restriction of Q to W_N when N is σ-integrable.

The next proposition is a slight modification of a result in [**DM**, XVIII-23].

(9.16) Proposition. *Let N be a* SM *and define*

$$K := \bigcap_n \{w\colon N(w,]-1/n, 1/n[) > 0\} .$$

Then P_N is carried by K.

Proof. Let h be the indicator of K^c. Given $t \in \mathbb{R}$ it follows that $h(\sigma_t w) = 0$ if and only if t is in the support of $N(w,\cdot)$. Consequently by (9.13)

$$P_N(K^c) = Q \int_0^1 h \circ \sigma_t N(dt) = 0 . \quad \blacksquare$$

(9.17) Remarks. When T is a ST we write P_T for the Palm measure, P_N, of $N := 1_{\mathbb{R}}(T)\epsilon_T$. It is immediate from the discussion surrounding (9.14) that P_T is σ-finite. Moreover, (9.16) implies that P_T is carried by $\{T = 0\}$. Using (9.11) with φ the indicator of $[0, a[$, we see that

$$P_T(f) = a^{-1} Q(f \circ \sigma_T; 0 \le T < a).$$

In particular if Q is a finite measure, then $Q(0 \le T < a) = a\, P_T(1)$ and so

$$P_T(f)/P_T(1) = \frac{Q(f \circ \sigma_T; 0 \le T < a)}{Q(0 \le T < a)}.$$

This leads to the interpretation of $P_T/P_T(1)$ as the "conditional law" given $T = 0$.

The next result is a "switching" identity that is due to Neveu [**N76**].

(9.18) Proposition. *Let N_1 and N_2 be* SMs. *Then for $F \in p(\mathcal{B} \otimes \mathcal{G}^0)$ one has*

$$P_{N_2} \int F(t, \cdot) N_1(dt) = P_{N_1} \int F(-t, \sigma_t) N_2(dt).$$

Proof. Let $f := \int F(t, \cdot) N_1(dt)$ and $\varphi \in p\mathcal{B}$ with $\ell(\varphi) = 1$. Then using (9.11)

$$\begin{aligned}
P_{N_2} \int F(t, \cdot) N_1(dt) &= Q \int \varphi(s) f \circ \sigma_s N_2(ds) \\
&= Q \iint \varphi(s) F(t - s, \sigma_s) N_2(ds) N_1(dt) \\
&= Q \int \left[\int \varphi(s + t) F(-s, \sigma_s) N_2(ds) \right] \circ \sigma_t N_1(dt) \\
&= \int dt\, P_{N_1} \int \varphi(s + t) F(-s, \sigma_s) N_2(ds) \\
&= P_{N_1} \int F(-s, \sigma_s) N_2(ds). \quad \blacksquare
\end{aligned}$$

Let $\varphi \in p\mathcal{B}$. Then (9.18) implies

$$P_{N_2}(N_1(\varphi)) = P_{N_1}(N_2(\hat{\varphi})) \tag{9.19}$$

where $\hat{\varphi}(t) = \varphi(-t)$. (Recall $N(\varphi) = \int \varphi(t)N(dt)$.) Combining this with (9.11) yields for $\psi \in p\mathcal{B}$ with $0 < \ell(\psi) < \infty$

$$\begin{aligned} Q \iint \psi(t)\varphi(s-t)N_1(ds)N_2(dt) \\ = Q \iint \psi(s)\varphi(s-t)N_1(ds)N_2(dt). \end{aligned} \tag{9.20}$$

This should be compared with (9.6). If S and T are STs, then (9.19) states that the distribution of S under P_T is the same as the distribution of $-T$ under P_S. Also (9.20) becomes

$$\begin{aligned} Q[\psi(T)\varphi(S-T);\ S \in \mathbb{R},\ T \in \mathbb{R}] \\ = Q[\psi(S)\varphi(S-T);\ S \in \mathbb{R},\ T \in \mathbb{R}]. \end{aligned}$$

But $1_{\mathbb{R}\times\mathbb{R}}(S,T)\varphi(S-T)$ is invariant and so this is just a special case of (9.8).

We record next yet another "switching" identity for Palm measures.

(9.21) Proposition. *Let N_1 and N_2 be* SMs. *Let $\Lambda := \mathbb{R}\times\mathbb{R}\times W \times W$ and $\mathcal{B}(\Lambda) = \mathcal{B}\otimes\mathcal{B}\otimes\mathcal{G}^0\otimes\mathcal{G}^0$. Define an s-finite measure γ on Λ by*

$$\gamma(H) = Q \iint H(s,t,\sigma_s,\sigma_t)N_1(ds)N_2(dt)$$

for $H \in p\mathcal{B}(\Lambda)$. Then there exists a unique s-finite measure $\mu = \mu_{N_1,N_2}$ on $\mathbb{R}\times W\times W$ such that

$$\begin{aligned} \gamma(H) &= \int ds \int H(s, s+t, w_1, w_2)\mu(dt,\, dw_1,\, dw_2) \\ &= \int ds \int H(s-t, s, w_1, w_2)\mu(dt,\, dw_1,\, dw_2), \end{aligned} \tag{9.22}$$

and one has

$$(9.23) \qquad \mu_{N_1,N_2}(dt,\, dw_1,\, dw_2) = \mu_{N_2,N_1}(-dt,\, dw_2,\, dw_1)\,.$$

Remark. The relationship (9.23) may be written more formally as follows: Let τ be the map from $\mathbb{R}\times W\times W$ to itself defined by $\tau(t,w_1,w_2) = (-t,w_2,w_1)$, then $\mu_{N_1,N_2} = \tau(\mu_{N_2,N_1})$.

Proof. Define an s-finite measure λ on Λ by

$$\lambda(H) = Q\iint H(s,t-s,\sigma_s,\sigma_t)N_1(ds)N_2(dt)$$

for $H \in p\mathcal{B}(\Lambda)$. If $u \in \mathbb{R}$, let $H_u(s,t,w_1,w_2) := H(s+u,t,w_1,w_2)$. Then

$$\begin{aligned}\lambda(H_u) &= Q\iint H(s+u,t+u-(s+u),\sigma_s,\sigma_t)N_1(ds)N_2(dt)\\ &= Q\left[\left[\iint H(s,t-s,\sigma_s,\sigma_t)N_1(ds)N_2(dt)\right]\circ\sigma_{-u}\right]\\ &= \lambda(H)\,,\end{aligned}$$

where the second equality uses the fact that N_1 and N_2 are SMs and the third equality uses the invariance of Q under σ_{-u}. Consequently by (8.23) there exists an s-finite measure μ on $\mathbb{R}\times W\times W$ such that

$$\lambda(H) = \int ds\int H(s,t,w_1,w_2)\mu(dt,\, dw_1,\, dw_2)\,.$$

Let $\tilde{H}(s,t,w_1,w_2) := H(s,s+t,w_1,w_2)$. Then $\gamma(H) = \lambda(\tilde{H})$ and combining this with the above gives the first equality in (9.22). The second equality in (9.22) then follows from the translation invariance of Lebesgue measure. If $h \in p(\mathcal{B} \otimes \mathcal{G}^0 \otimes \mathcal{G}^0)$ and $\varphi \in p\mathcal{B}$ with $\ell(\varphi) = 1$, then setting $H(s,t,\cdot,\cdot) := \varphi(s)h(t-s,\cdot,\cdot)$, (9.22) becomes

$$\mu(h) = Q \iint \varphi(s)h(t-s,\sigma_s,\sigma_t)N_1(ds)N_2(dt)\,.$$

In particular μ is unique.

For (9.23) let $\mu = \mu_{N_1,N_2}$ and $\nu = \mu_{N_2,N_1}$. Also write γ_{N_1,N_2} for γ as defined in (9.21) and γ_{N_2,N_1} for the corresponding measure with N_1 and N_2 interchanged. Finally define $\hat{H}(s,t,w_1,w_2) := H(t,s,w_2,w_1)$ for $H \in p\mathcal{B}(\Lambda)$. Then

$$\begin{aligned}
\gamma_{N_1,N_2}(H) &= \gamma_{N_2,N_1}(\hat{H}) \\
&= \int ds \int \hat{H}(s,s+t,w_1,w_2)\nu(dt,\,dw_1,\,dw_2) \\
&= \int ds \int H(s+t,s,w_2,w_1)\nu(dt,\,dw_1,\,dw_2) \\
&= \int ds \int H(s,s-t,w_2,w_1)\nu(dt,\,dw_1,\,dw_2) \\
&= \int ds \int H(s,s+t,w_1,w_2)\nu(-dt,\,dw_2,\,dw_1)\,.
\end{aligned}$$

Combining this with (9.22) gives (9.23). ■

There are many variations on the theme of Proposition 9.21. We content ourselves with just one.

(9.24) **Proposition.** *Let N_1 and N_2 be* SM*s and let $a, b \in \mathbb{R}$ with $a + b \neq 0$. Then for $h \in p(\mathcal{B} \otimes \mathcal{B})$*

$$\begin{aligned}
&Q \iint h(s,t)N_1(ds)N_2(t) \\
&\quad = |a+b|Q \iint h(as+bt,(a-1)s+(b+1)t)N_1(ds)N_2(dt)\,.
\end{aligned}$$

Proof. Let $\varphi \in p\mathcal{B}$ with $\ell(\varphi) = 1$. Let J denote the left hand side of the formula in (9.24). Then using the stationarity of N_1 and N_2 and the invariance of Q

$$J = \int \varphi(u) J \, du$$
$$= \int \varphi(u) Q \left[\iint h(s-u, t-u) N_1(ds) N_2(dt) \right] du \,.$$

Interchanging the order of integration in u and w and making the change of variable $u = (1-a)s - bt + (a+b)v$ this becomes

$$J = |a+b| Q \iiint \varphi[(1-a)(s-v) - b(t-v) + v]$$
$$\times h[a(s-v) + b(t-v), (a-1)(s-v) + (b+1)(t-v)] N_1(ds) N_2(dt) \, dv$$
$$= |a+b| Q \iiint \varphi[(1-a)s - bt + v) \, dv$$
$$\times \; h(as+bt, (a-1)s + (b+1)t) N_1(ds) N_2(dt)$$
$$= |a+b| Q \iint h(as+bt, (a-1)s + (b+1)t) N_1(ds) N_2(dt) \,.$$

(9.25) **Remarks.** If S and T are STs, (9.24) becomes

$$Q[h(S,T);\; S \in \mathbb{R};\; T \in \mathbb{R}]$$
$$= |a+b| Q[h(aS+bT, (a-1)S + (b+1)T);\; S \in \mathbb{R},\; T \in \mathbb{R}] \,.$$

Specializing to $a = 0$ and $b = -1$ gives

$$Q[h(S,T);\; S \in \mathbb{R};\; T \in \mathbb{R}] = Q[h(-T,-S);\; S \in \mathbb{R};\; T \in \mathbb{R}] \,.$$

Taking $h(s,t) = \varphi(s)$ we obtain for $a + b \neq 0$

$$Q[\varphi(S);\; S \in \mathbb{R};\; T \in \mathbb{R}] = |a+b| Q[\varphi(aS+bT);\; S \in \mathbb{R};\; T \in \mathbb{R}] \,.$$

In particular if $|a+b|=1$, the distribution of $aS+bT$ on $\{S \in \mathbb{R},\ T \in \mathbb{R}\}$ is independent of a and b. If $h(s,t)=\varphi(t-s)$, we find

$$Q[\varphi(T-S);\ S \in \mathbb{R};\ T \in \mathbb{R}] \\ = |a+b| Q[\varphi(T-S);\ S \in \mathbb{R};\ T \in \mathbb{R}],$$

so that the integral on the left hand side of this last display is either zero or infinity. More generally if $J \in p\mathcal{I}$ and S is a ST, then

$$Q(J;\ S \in \mathbb{R}) = \sum_{n \in \mathbb{Z}} Q(J;\ n \le S < n+1) \\ = \sum_{n \in \mathbb{Z}} Q(J;\ 0 \le S < 1),$$

which is zero or infinity according as $Q(J;\ 0 \le S < 1)$ is zero or nonzero.

We shall next give a characterization of the Palm measure of a SM. A set $\Lambda \in \mathcal{G}^0$ is σ-Q *polar* provided $T_\Lambda(w) := \inf\{t \in \mathbb{R}: \sigma_t w \in \Lambda\}$ satisfies $Q(T_\Lambda < \infty) = 0$. As usual the infimum of the empty set is infinity. If $g = 1_\Lambda$, then Λ is σ-Q polar if and only if the process $(g \circ \sigma_t)_{t \in \mathbb{R}}$ is Q-evanescent. It is obvious from (9.11) or (9.13), that if N is a SM, then P_N does not charge σ-Q polars. We are going to show that this, in fact, characterizes the Palm measure of a SM. We prepare several lemmas for the proof.

First of all we say that an s-finite measure, μ, on $\mathbb{R} \times W$ *respects* Q provided given $f \in p\mathcal{G}^0$ with $f = 0$ a.s. Q, then for $\varphi \in p\mathcal{B}$, $\varphi \otimes f = 0$ a.e. μ. Here $\varphi \otimes f(t,w) := \varphi(t) f(w)$.

(9.26) Lemma. *Let P be an s-finite measure on $(W, \mathcal{G}^0)$ not charging σ-Q-polars. Let $\rho: (t,w) \to (t, \sigma_{-t} w)$ map $\mathbb{R} \times W$ onto itself. Then $\rho(\ell \times P)$ respects Q.*

Proof. Obviously $\rho \in \mathcal{B} \otimes \mathcal{G}^0 | \mathcal{B} \otimes \mathcal{G}^0$ and, of course, ℓ denotes Lebesgue measure on $\mathbb{R}$. Let $f \in pb\mathcal{G}^0$ with $f = 0$ a.s. Q.

Let $\psi \in p\mathcal{B}$ be continuous with compact support. Define $Y := \int \psi(t) f \circ \sigma_t \, dt$. Now

$$Y \circ \sigma_s = \int \psi(t) f \circ \sigma_{t+s} \, dt = \int \psi(t-s) f \circ \sigma_t \, dt ,$$

and $Q(f \circ \sigma_t) = Q(f) = 0$. Hence by the Fubini theorem $Y \circ \sigma_s = 0$ a.s. Q for each s. But $s \to Y \circ \sigma_s$ is continuous and so $Y \circ \sigma_s$ is Q evanescent. Therefore $P(Y) = 0$ and consequently $(t, w) \to f(\sigma_t w)$ vanishes a.e. $\ell \times P$. Let $\nu = \rho(\ell \times P)$ and $\varphi \in p\mathcal{B}$. Then

$$\nu(\varphi \otimes f) = \int \varphi(t) P(f \circ \sigma_{-t}) \, dt = P \int \varphi(-t) f \circ \sigma_t \, dt = 0 . \quad \blacksquare$$

The next lemma is essentially [**S**, (30.4)].

(9.27) Lemma. *Let μ be an s-finite measure on $(\mathbb{R} \times W, \mathcal{B} \otimes \mathcal{G}^0)$ that respects Q. Then there exists a kernel $N(w, dt)$ from $(W, \mathcal{G}^0)$ to $(\mathbb{R}, \mathcal{B})$ that is a countable sum of bounded kernels from $(W, \mathcal{G}^0)$ to $(\mathbb{R}, \mathcal{B})$ such that $\mu(F) = Q \int F(t, \cdot) N(dt)$ for $F \in p(\mathcal{B} \otimes \mathcal{G}^0)$. If μ is finite, N is integrable; that is, $Q[N(\mathbb{R})] < \infty$. If μ is σ-finite, N is σ-integrable and unique.*

Remark. The uniqueness assertion means that if N_1 and N_2 are two such kernels then $N_1(w, \cdot) = N_2(w, \cdot)$ as measures on $\mathbb{R}$ for Q a.e. w. N is σ-integrable provided there exists $H \in \mathcal{B} \otimes \mathcal{G}^0$ with $H > 0$ and $Q \int H(t, \cdot) N(dt) < \infty$.

Proof. Suppose first that μ is finite. Given $\varphi \in b\mathcal{B}$, define μ^φ on $(W, \mathcal{G}^0)$ by $\mu^\varphi(f) := \mu(\varphi \otimes f)$ for $f \in b\mathcal{G}^0$. Then μ^φ is a signed measure on $(W, \mathcal{G}^0)$. If $f \in pb\mathcal{G}^0$ and $f = 0$ a.s. Q, then because μ respects Q, $\mu^\varphi(f) = 0$. Consequently $\mu^\varphi \ll Q$. In particular $\mu^1 = f_0 Q$ where $f_0 \in p\mathcal{G}^0$ satisfies $0 \le f_0 < \infty$ and $Q(f_0) = \mu^1(1) = \mu(1) < \infty$. Clearly $\mu^\varphi \ll \mu^1$ if $\varphi \in b\mathcal{B}$, and so

$$\mu^\varphi(f) = \int f(w) n(w, \varphi) \mu^1(dw)$$

for $f \in b\mathcal{G}^0$ where $n(\cdot,\varphi)$ is a version of $d\mu^\varphi/d\mu^1$. It is evident that $\varphi \to n(\cdot,\varphi)$ is a subMarkovian pseudo-kernel from W to $\mathbb{R}$ relative to $\mathcal{N} := \{\Lambda \in \mathcal{G}^0 : \mu^1(\Lambda) = 0\}$. See [**DM**, IX-11]. Consequently by the theorem in [**DM**, IX-11], there exists a subMarkovian kernel $\bar{N}$ from $(W,\mathcal{G}^0)$ to $(\mathbb{R},\mathcal{B})$ such that $n(\cdot,\varphi) = \bar{N}(\varphi)$ a.e. μ^1 for each $\varphi \in b\mathcal{B}$. Define $N(w,dt) := f_0(w)\bar{N}(w,dt)$. Then N is a finite kernel from $(W,\mathcal{G}^0)$ to $(\mathbb{R},\mathcal{B})$ such that

$$\mu(\varphi \otimes f) = \int f(w)N(w,\varphi)Q(dw)$$

for $\varphi \in b\mathcal{B}$ and $f \in b\mathcal{G}^0$. Moreover $Q[N(\mathbb{R})] = \mu(1) < \infty$. Consequently the monotone class theorem implies that $\mu(F) = Q\int F(t,\cdot)N(dt)$ for $F \in b(\mathcal{B} \otimes \mathcal{G}^0)$. If μ is s-finite, write $\mu = \sum \mu_k$ where each μ_k is finite and set $N = \sum N_k$ where each N_k is constructed from μ_k as above. This establishes the existence of N for a general s-finite μ respecting Q. Suppose μ is σ-finite and let $H \in \mathcal{B} \otimes \mathcal{G}^0$ with $H > 0$ and $\mu(H) < \infty$. Then $Q\int H(t,\cdot)N(dt) = \mu(H) < \infty$ and so N is σ-integrable. If M is another such kernel, then, using the separability of $\mathcal{B}$, it is easy to check that $H(\cdot,w)N(w,\cdot) = H(\cdot,w)M(w,\cdot)$ as finite measures on $\mathbb{R}$ for Q a.e. w, and since $H > 0$, this establishes the uniqueness of N when μ is σ-finite. ∎

Remark. Obviously the conclusion of (9.27),

$$\mu(F) = Q\int F(t,\cdot)N(dt),$$

extends to $F \in p(\mathcal{B} \otimes \mathcal{G})$ where $\mathcal{G}$ is the Q-completion of $\mathcal{G}^0$, or, more generally, to $F \geq 0$ and measurable relative to the μ-completion of $\mathcal{B} \otimes \mathcal{G}^0$.

We next formulate and prove a preliminary version of the perfection theorem for SMs alluded to in (9.5). The argument comes from [**FS89**]. The result will be improved in (9.34) and (9.35).

(9.28) Lemma. *Let K be a kernel from $(W, \mathcal{G}^0)$ to $(\mathbb{R}, \mathcal{B})$ that satisfies:*

$$\text{(9.29)} \qquad Q \int \varphi(t) K(dt) < \infty \quad \textit{for } \varphi \in p b\mathcal{B} \textit{ with } \ell(\varphi) < \infty;$$

$$\text{(9.30)} \qquad K(\sigma_s w, B) = K(w, B+s) \quad \textit{for} \quad Q \text{ a.e. } w$$

for each fixed $s \in \mathbb{R}$ and $B \in \mathcal{B}$. Then there exists a SM, *N, which is equivalent to K; that is, $N(w, \cdot) = K(w, \cdot)$ as measures on $(\mathbb{R}, \mathcal{B})$ for Q a.e. w. Moreover, N may be chosen to be a kernel from $(W, \mathcal{G}^0)$ to $(\mathbb{R}, \mathcal{B})$.*

Proof. Define $K_s(w, B) = K(\sigma_s w, B - s)$. The hypothesis (9.30) implies that $K_s(\cdot, B) = K(\cdot, B)$ a.s. for each $s \in \mathbb{R}$ and $B \in \mathcal{B}$. Moreover if $\varphi \in p b\mathcal{B}$ with $\ell(\varphi) < \infty$, then

$$\begin{aligned} Q \int \varphi(t) K_s(dt) &= Q \int \varphi(t+s) K(\sigma_s, dt) \\ &= Q \int \varphi(t+s) K(dt) < \infty, \end{aligned}$$

where the second equality uses $\sigma_s(Q) = Q$. Let $\mathcal{B}_0$ be the collection of compact intervals with rational endpoints. Define

$$D := \{(s, w) : K_s(w, J) = K(w, J) < \infty; \ \forall J \in \mathcal{B}_0\}.$$

Then $D \in \mathcal{B} \otimes \mathcal{G}^0$ and the hypotheses imply that $\ell \times Q(D^c) = 0$. If $(s, w) \in D$, then $K_s(w, \cdot)$ and $K(w, \cdot)$ are Radon measures which agree on $\mathcal{B}_0$, and so $K_s(w, \cdot) = K(w, \cdot)$. Next define

$$\begin{aligned} \Gamma :&= \{w : (s, w) \in D \text{ for a.e. } s\} \\ &= \left\{ w : \int 1_{D^c}(s, w)\, ds = 0 \right\} \in \mathcal{G}^0. \end{aligned}$$

Then $Q(\Gamma^c) = 0$ and $K_s(w, \cdot) = K(w, \cdot)$ for a.e. s if $w \in \Gamma$. Finally define

$$\Lambda := \{w : \sigma_s w \in \Gamma \text{ for } a.e.s\}$$
$$= \left\{ w : \int 1_{\Gamma^c}(\sigma_s w)\, ds = 0 \right\} \in \mathcal{G}^0 .$$

It is evident that $\Lambda = \sigma_t^{-1}(\Lambda)$ for each t. Because $Q[\sigma_t^{-1}(\Gamma^c)] = 0$ for each t, the Fubini theorem implies that $Q(\Lambda^c) = 0$.

Now let $\Phi := \{\varphi \in pb\mathcal{B} : \ell(\varphi) = 1\}$. Given $\varphi \in \Phi$, define

$$K^\varphi(w, B) := \int \varphi(s) K_s(w, B)\, ds = \int \varphi(s) K(\sigma_s w, B - s)\, ds .$$

It is immediate that for each $\varphi \in \Phi$, K^φ is a kernel from $(W, \mathcal{G}^0)$ to $(\mathbb{R}, \mathcal{B})$ and that $K^\varphi(w, \cdot) = K(w, \cdot)$ for $w \in \Gamma$. If $\varphi \in \Phi$, let $\varphi_s(t) := \varphi(t - s)$. For $w \in \Lambda$,

$$K^{\varphi_s}(w, B) = \int \varphi(t - s) K(\sigma_t w, B - t)\, dt$$
$$= \int \varphi(t) K(\sigma_{s+t} w,\ B - t - s)\, dt .$$

But $\sigma_t w \in \Gamma$ for a.e. t because $w \in \Lambda$, and so for a.e. t, $K(\sigma_s \sigma_t w, B - t - s) = K(\sigma_t w, B - t)$ for a.e. s. Another application of the Fubini theorem yields $K^{\varphi_s}(w, B) = K^\varphi(w, B)$ for a.e. s. Now given $\psi \in \Phi$,

$$K^\varphi(w, B) = \int \psi(s) K^{\varphi_s}(w, B)\, ds$$
$$= \int \psi(s) \int \varphi(t) K(\sigma_{t+s} w, B - t - s)\, dt\, ds$$
$$= \int \varphi(t) K^{\psi_t}(w, B)\, dt = K^\psi(w, B) .$$

Therefore $K^{\varphi}(w,\cdot) = K^{\psi}(w,\cdot)$ for $w \in \Lambda$ and $\varphi, \psi \in \Phi$. Define $N(w,\cdot) = 1_{\Lambda}(w)K^{\varphi}(w,\cdot)$. By the above this is independent of the choice of $\varphi \in \Phi$ and $N(w,\cdot) = K(w,\cdot)$ for Q a.e. w. In addition N is a kernel from $(W, \mathcal{G}^0)$ to $(\mathbb{R}, \mathcal{B})$. Given $t \in \mathbb{R}$, $w \in W$, and $B \in \mathcal{B}$ one has

$$\begin{aligned} N(\sigma_t w, B) &= 1_{\Lambda}(\sigma_t w) \int \varphi(s) K(\sigma_{s+t} w, B - s)\, ds \\ &= 1_{\Lambda}(w) \int \varphi(s-t) K(\sigma_s w, B + t - s)\, ds \\ &= 1_{\Lambda}(w) K^{\varphi_t}(w, B+t) = N(w, B+t), \end{aligned}$$

where the second equality uses the invariance of Λ and the fourth holds because $\varphi_t \in \Phi$. ∎

At long last we come to the promised characterization of Palm measures.

(9.31) Theorem. *An s-finite measure P on $(W, \mathcal{G}^0)$ is the Palm measure of a* SM, *N, if and only if P does not charge σ-Q polars. In fact, N may be assumed to be a kernel from $(W, \mathcal{G}^0)$ to $(\mathbb{R}, \mathcal{B})$. If P is σ-finite, then N is σ-integrable and is unique up to equivalence.*

Proof. We have already pointed out that P_N does not charge σ-Q polars when N is a SM. It suffices to prove the existence of N under the additional assumption that P is finite. Let $\rho\colon (t,w) \to (t, \sigma_{-t} w)$ as in (9.26) and let $\mu = \rho(\ell \times P)$. By (9.26), μ respects Q. Obviously μ is s-finite and so, according to (9.27), there exists a kernel N from $(W, \mathcal{G}^0)$ to $(\mathbb{R}, \mathcal{B})$ such that

$$Q \int F(t,\cdot) N(dt) = \mu(F) = P \int F(t, \sigma_{-t})\, dt \tag{9.32}$$

for $F \in p(\mathcal{B} \otimes \mathcal{G}^0)$. Let $\varphi \in p\mathcal{B}$ with $\ell(\varphi) < \infty$. Then (9.32) implies that $Q \int \varphi(t) N(dt) = \ell(\varphi) P(1) < \infty$. Consequently

N satisfies (9.29). We claim that N also satisfies (9.30). To this end let $f \in p\mathcal{G}^0$ and $\varphi \in p\mathcal{B}$. Then for $s \in \mathbb{R}$ fixed

$$\begin{aligned} Q\left[f\int \varphi(t-s)N(dt)\right] &= \int dt\,\varphi(t-s)P(f\circ\sigma_{-t}) \\ &= \int dt\,\varphi(t)P(f\circ\sigma_{-t-s}) \\ &= Q\left[f\circ\sigma_{-s}\int \varphi(t)N(dt)\right] \\ &= Q\left[f\int \varphi(t)N(\sigma_s, dt)\right], \end{aligned}$$

where the first and third equalities come from (9.32) and the last uses $\sigma_s(Q) = Q$. It is now clear that $N(\sigma_s w, B) = N(w, B+s)$ for Q a.e. w for each fixed $s \in \mathbb{R}$ and $B \in \mathcal{B}$. Therefore, because of (9.28), we may suppose that N is, in fact, a SM for which (9.32) holds, and that N is a kernel from $(W, \mathcal{G}^0)$ to $(\mathbb{R}, \mathcal{B})$. Replacing $F(t,w)$ by $F(t, \sigma_t w)$ in (9.32), ones sees that $P = P_N$. This establishes the existence of N.

It was shown in the discussion following (9.13) that N is σ-integrable if and only if $P_N = P$ is σ-finite. In this case choose $h > 0$ with $P(h) < \infty$ and $\varphi > 0$ with $\ell(\varphi) < \infty$. Then $H(t,\cdot) := \varphi(t)h\circ\sigma_t > 0$, and by (9.32),

$$Q\int H(t,\cdot)N(dt) = \ell(\varphi)P(h) < \infty.$$

Hence $H(t,\cdot)N(dt)$ is a finite measure a.s., and (9.32) implies that it is uniquely determined up to equivalence by P. Consequently N is unique up to equivalence. ∎

We turn now to a perfection theorem for (weakly) stationary measures. Let τ_s be the translation operator on $\mathbb{R}$ defined by $\tau_s(t) = t - s$. A *weakly stationary measure* (WSM) is a kernel, K, from $(W, \mathcal{G})$ to $(\mathbb{R}, \mathcal{B})$ that is a countable sum of bounded kernels and such that for each $s \in \mathbb{R}$

$$(9.33) \qquad K(\sigma_s w, \cdot) = \tau_s(K(w, \cdot)) \quad \text{for} \quad Q \text{ a.e. } w.$$

Of course, the exceptional is allowed to depend on s. A WSM, K, is σ-integrable provided there exists $H \in p(\mathcal{B} \otimes \mathcal{G})$ with $H > 0$ and $Q \int H(t, \cdot) K(dt) < \infty$. Then there exists such an $H \in p(\mathcal{B} \otimes \mathcal{G}^0)$.

(9.34) Theorem. *Let K be a σ-integrable* WSM. *Then there exists a* SM, N, *equivalent to* K.

Proof. Given any WSM, K, arguing exactly as in the proof of (9.9), there exists a unique s-finite measure P_K—the Palm measure of K—on $(W, \mathcal{G})$ such that

$$Q \int F(t, \sigma_t) K(dt) = \int dt \, P_K(F(t, \cdot))$$

for $F \in p(\mathcal{B} \otimes \mathcal{G})$. Clearly P_K does not charge σ-Q polars. Hence by (9.31)—see also (9.32)—there exists a SM, N, such that

$$Q \int F(t, \ ) K(dt) = Q \int F(t, \cdot) N(dt).$$

for $F \in p(\mathcal{B} \otimes \mathcal{G})$. If K is σ-integrable, then one easily checks that N and K are equivalent. ∎

(9.35) Remarks. It is natural to ask whether (9.33) may be replaced by the weaker condition (9.30) in the above theorem. Obviously this may be done whenever (9.30) implies (9.33). The argument in the first part of the proof of (9.28) shows that this is possible provided $K(w, \cdot)$ is a Radon measure for a.e. w. It is reasonable to conjecture that if there exists $f \in p\mathcal{G}^0$ with $f > 0$ and $Q \int f \circ \sigma_t K(dt) < \infty$, then one should be able to "perfect" the finite measure $M(dt) := f \circ \sigma_t K(dt)$. The problem is that it is not clear to me that M satisfies (9.30) when K does. Alternatively one might try to prove the existence of P_K when K satisfies (9.30). But this involves showing that the measure η defined by the left hand side of (9.10) with N replaced by K, is translation invariant along $\mathbb{R}$. Defining $K_s(w, B) = K(\sigma_s w, B - s)$, this will follow provided

$$(9.36) \qquad Q \int F(t, \cdot) K_s(dt) = Q \int F(t, \cdot) K(dt)$$

for $F \in p(\mathcal{B} \otimes \mathcal{G}^0)$. This is clear for $F = \varphi \otimes f$, and the measures defined by the two sides of (9.36) are σ-finite if K is σ-integrable. But, of course, this is not enough to conclude that (9.36) holds.

We close this section by discussing the decomposition of Q into "dissipative" and "conservative" parts following **[F88a]**. The results are analogous to, but simpler than, those detailed in (6.24) and (6.25). We fix $(W, \mathcal{G}^0, (\sigma_t)_{t \in \mathbb{R}})$ satisfying (9.1i) and (9.1ii) and we write $\mathcal{Q}(\sigma)$ for the family of σ-finite measures Q on $(W, \mathcal{G}^0)$ which are invariant under the group $(\sigma_t)_{t \in \mathbb{R}}$. Thus for each $Q \in \mathcal{Q}(\sigma)$, $(W, \mathcal{G}^0, Q, (\sigma_t)_{t \in \mathbb{R}})$ is a flow as defined in (9.1). Recall that for $F \in p\mathcal{G}^0$, $\bar{F} := \int_{-\infty}^{\infty} F \circ \sigma_t \, dt$.

(9.37) Definition. $Q \in \mathcal{Q}(\sigma)$ is *dissipative* if $\bar{F} < \infty$ a.s. Q for every $F \in p\mathcal{G}^0$ with $Q(F) < \infty$, and Q is *conservative* if $\bar{F} = \infty$ a.s. Q for each strictly postive $F \in \mathcal{G}^0$.

(9.38) Proposition. *Let $Q \in \mathcal{Q}(\sigma)$. (i) Q is dissipative if and only if there exists a stationary time S with $Q(S \notin \mathbb{R}) = 0$. (ii) Q is conservative if and only if $Q(S \in \mathbb{R}) = 0$ for every stationary time S.*

Proof. (i) Suppose Q is dissipative. Choose $F \in \mathcal{G}^0$ with $F > 0$ and $Q(F) < \infty$. Then $Q(\bar{F} = \infty) = 0$. Consequently a.s. Q, $A_t := \int_{-\infty}^{t} F \circ \sigma_s \, ds$ is strictly positive, continuous, and decreases to zero as $t \downarrow -\infty$. For $n \geq 1$, let $S_n = \inf\{t: A_t > 1/n\}$. It is easily verified that each S_n is a ST. Moreover a.s. Q, $S_n > -\infty$ and $S_n \downarrow -\infty$ as $n \to \infty$. Let $S_0 := +\infty$ and define S by

$$\begin{aligned} S &= S_n \quad \text{on} \quad \{\bar{F} < \infty\} \cap \{S_n < \infty,\, S_{n-1} = \infty\} \\ &= \infty \quad \text{on} \quad \{\bar{F} = \infty\} \cup \{S_n = \infty \quad \text{for every} \quad n\}. \end{aligned}$$

It is clear that S is a ST (recall $\bar{F} \in \mathcal{I}$) and that $Q(S \notin \mathbb{R}) = 0$ because $Q(\cap\{S_n = \infty\}) = 0$. Conversely suppose S is a ST with $Q(S \notin \mathbb{R}) = 0$. Let $F \in p\mathcal{G}^0$ with $Q(F) < \infty$. Then using (9.2) and the fact that for $\varphi \in p\mathcal{B}$, $\varphi(S)^- = \ell(\varphi)\, 1_{\mathbb{R}}(S)$ one finds

$$Q(\bar{F}\varphi(S)) = \ell(\varphi) Q(F;\, S \in \mathbb{R}).$$

Choosing $\varphi > 0$ with $\ell(\varphi) < \infty$, it follows that $\bar{F} < \infty$ a.s. Q on $\{S \in \mathbb{R}\}$. Hence $Q(\bar{F} = \infty) = 0$ and so Q is dissipative.

(ii) Suppose Q is conservative and that S is a stationary time with $Q(S \in \mathbb{R}) > 0$. Then there exist $-\infty < a < b < \infty$ with $Q(a < S < b) > 0$. Let $F > 0$ with $Q(F) < \infty$. Then

$$\infty > (b-a) Q(F;\, S \in \mathbb{R}) = Q(\bar{F};\, a < S < b)$$

which contradicts the fact that $\bar{F} = \infty$ a.s. Q. Conversely suppose $Q(S \in \mathbb{R}) = 0$ for all stationary times and that there exists $F > 0$ with $Q(F) < \infty$ and $Q(\bar{F} < \infty) > 0$. As in (i) let $A_t = \int_{-\infty}^{t} F \circ \sigma_s \, ds$. Since $A_\infty = \bar{F}$, $Q(0 < A_\infty < \infty) > 0$. Therefore $S_n := \inf\{t: A_t > 1/n\}$ is a ST and for n sufficiently large $Q(S_n \in \mathbb{R}) > 0$. ∎

(9.39) Theorem. *Let $Q \in \mathcal{Q}(\sigma)$. Then Q may be written uniquely as $Q = Q_c + Q_d$ where Q_c is conservative and Q_d is dissipative.*

Proof. Fix $F > 0$ with $Q(F) < \infty$. Let $f = 1_{\{\bar{F}=\infty\}}$ and $g = 1 - f = 1_{\{\bar{F}<\infty\}}$. Define $Q_c := fQ$ and $Q_d := gQ$. Since

f and g are invariant both Q_c and Q_d are in $\mathcal{Q}(\sigma)$. Suppose there exists a ST, S, with $Q_c(S \in \mathbb{R}) > 0$. Then there exists a nonvoid bounded interval $]a,b[$ with $Q(f;\, a < S < b) = Q_c(a < S < b) > 0$. Therefore

$$\infty > Q_c(F;\, S \in \mathbb{R}) = (b-a)Q_c(\bar{F};\, a < S < b) \\ = (b-a)Q(f\bar{F};\, a < S < b) = \infty\,,$$

and so Q_c is conservative by (9.38). With F as above let S be defined as in the proof of (9.38i). Since Q_d is carried by $\{\bar{F} < \infty\}$, it follows that $Q_d(S \notin \mathbb{R}) = 0$ and so Q_d is dissipative. It remains to prove the uniqueness of this decomposition. Suppose $Q = Q'_c + Q'_d$ where Q'_c is conservative and Q'_d is dissipative. Choose a ST, S, with $Q_d(S \notin \mathbb{R}) = 0$. For $H \in p\mathcal{G}^0$ one then has because $Q_c(S \in \mathbb{R}) = 0 = Q'_c(S \in \mathbb{R})$,

$$Q_d(H) = Q_d(H;\, S \in \mathbb{R}) = Q'_d(H;\, S \in \mathbb{R}) \le Q'_d(H)\,.$$

Consequently $Q_d \le Q'_d$ and by symmetry $Q_d = Q'_d$. It then follows that $Q_c = Q'_c$. ∎

10. Palm Measures and Capacity

Throughout this section X is a right process satisfying (6.2) and the notation of §6 is in force. In particular, for each $\xi \in \mathrm{Exc}$, $(W, \mathcal{G}^0, Q_\xi, (\sigma_t)_{t\in\mathbb{R}})$ is the Kuznetsov flow of ξ. If K is a HRM of Y as defined in (8.19), then it is a SM relative to each Kuznetsov flow and we write P_K^ξ for its Palm measure relative to Q_ξ. If $\nu = \nu_K^\xi$ denotes the Revuz measure of K defined in (8.25), then it is immediate that ν_K^ξ is just the restriction of $Y_0(P_K^\xi)$ to E. In particular, if N is a HRM of X, then the Revuz measure of N relative to ξ, ν_N^ξ is the restriction of $Y_0(P_{N^*}^\xi)$ to E where N^* is the extension of N to W defined in (8.18). The formulas for SMs developed in §9 may now be interpreted in the present set-up. For example, (9.18) implies

$$(10.1)\quad Q_\xi \int \varphi(s) \int F(t-s, Y_s) K_1(dt) K_2(ds)$$
$$= Q_\xi \int \varphi(t) \int F(t-s, Y_s) K_1(dt) K_2(ds)$$

for each $\xi \in \mathrm{Exc}$, K_1 and K_2 HRMs of Y, $\varphi \in p\mathcal{B}$, and $F \in p(\mathcal{B} \otimes \mathcal{G}^*)$.

Let N be a HRM of X and, for clarity, we continue to write N^* for the extension of N to W defined in (8.18). We define an s-finite measure $\mathbb{P}_N^\xi$ on $(W, \mathcal{G}^0)$ by

$$(10.2)\qquad \mathbb{P}_N^\xi(F) := P_{N^*}^\xi(F \circ b_0) = Q_\xi \int_0^1 F \circ \theta_t N^*(dt)\,.$$

We shall call $\mathbb{P}_N^\xi$ the *Palm measure of N relative to ξ*. Recall that N^* is carried by $]\alpha, \beta]$. Using the facts that $\alpha \circ \theta_t = 0$

and $Y_{\alpha+} \circ \theta_t = Y_t$ provided $\alpha < t < \beta$, one readily checks that $\mathbb{P}^{\xi}_N$ is carried by Ω, and so we may and shall regard $\mathbb{P}^{\xi}_N$ as a measure on $(\Omega, \mathcal{F}^0)$. If N is carried by $]0, \zeta[$, then N^* is carried by $]\alpha, \beta[$, and in this case $\nu^{\xi}_N = X_0(\mathbb{P}^{\xi}_N)$. In many ways, when dealing with X, $\mathbb{P}^{\xi}_N$ is a more natural object that $P^{\xi}_{N^*}$; but both are of importance.

Before proceeding we give some alternate expressions for the Palm measure under various assumptions on ξ. Suppose first that $\xi \in \mathrm{Pur}$ and that $\xi = \int_0^{\infty} \nu_t \, dt$ is the representation of ξ as the integral of an entrance law, $\nu = (\nu_t; t > 0)$. Let K be a HRM of Y, $F \in p\mathcal{G}^0$, and $\varphi \in p\mathcal{B}$ with $\ell(\varphi) = 1$. Then using (9.11) and (6.11) we have

$$
\begin{aligned}
(10.3) \qquad P^{\xi}_K(F) &= Q_{\xi} \int \varphi(t) F \circ \sigma_t K(dt) \\
&= \int ds \, Q_{\nu} \int \varphi(t) F \circ \sigma_{t+s} K(dt + s) \\
&= Q_{\nu} \int ds \int \varphi(t - s) F \circ \sigma_t K(dt) \\
&= Q_{\nu} \int_{[0,\infty[} F \circ \sigma_t K(dt),
\end{aligned}
$$

where for the last equality we have used the facts that $\alpha = 0$ a.s. Q_{ν} and that K is carried by $[\alpha, \beta]$. If N is a HRM of X, then N^* is carried by $]\alpha, \beta]$ and so

$$
\begin{aligned}
(10.4) \quad \mathbb{P}^{\xi}_N(F) = P^{\xi}_{N^*}(F \circ b_0) &= Q_{\nu} \int_{]0,\infty[} F \circ \theta_t N^*(dt) \\
&= \lim_{s \downarrow 0} Q_{\nu} \int_s^{\infty} F \circ \theta_t N^*(dt).
\end{aligned}
$$

If $\xi = \mu U$, it follows from (6.19) that Q_ν is carried by Ω and that $Q_\nu = P^\mu$ on Ω. Since $B \subset]s, \infty[$ and $\omega \in \Omega$ imply that $N^*(B) = N(\theta_s, B - s) = N(B)$, (10.4) becomes

$$\text{(10.5)} \qquad \mathbb{P}_N^{\mu U}(F) = P^\mu \int_0^\infty F \circ \theta_t N(dt),$$

for $F \in p\mathcal{F}^0$. Taking $F = f \circ X_0$, (10.5) reduces to a special case of (8.11). Next suppose $\xi \in \text{Inv}$. If N is a HRM of X and $F \in p\mathcal{F}^0$, then using (9.12), (10.2), and the fact that $\alpha = -\infty$ a.s. Q_ξ,

$$\text{(10.6)} \qquad \begin{aligned} \mathbb{P}_N^\xi(F) &= Q_\xi \int_0^1 F \circ \theta_t N^*(dt) \\ &= Q_\xi \left(\int_0^1 F \circ \theta_t N(dt) \circ \theta_0 \right) \\ &= P^\xi \int_0^1 F \circ \theta_t N(dt). \end{aligned}$$

Now let N be a HRM of X and $\psi : [0, \infty] \to [0, \infty]$ be continuous and increasing. Define $H = \psi[N(\mathbb{R}^+)]$. Then $H \circ \theta_t = \psi[N(]t, \infty])]$ and it follows that H is excessive as defined in §7. Note also that $H^* := \lim_{t \downarrow \alpha} H \circ \theta_t$ defined in (7.1) is given by $H^* = \psi[N^*(\mathbb{R})]$ where N^* is the extension of N defined in (8.18). Using this notation we may now formulate our next result which is taken from [**F88a**].

(10.7) Proposition. *With N, ψ, and H as above let $h(x) = P^x(H)$. Let $\xi \in \text{Dis}$ and S be a ST with $\alpha < S < \beta$ a.s. Q_ξ. If $\varphi \in p\mathcal{B}$ with $\ell(\varphi) = 1$, then $L(\xi, h) = Q_\xi(\varphi(S) H^*)$. If $\psi(t) = t$ so that $h = U_N 1$ one also has*

$$\text{(10.8)} \qquad L(\xi, U_N 1) = Q_\xi \int \varphi(t) N^*(dt) = Q_\xi(\varphi(S) N^*(\mathbb{R})).$$

Proof. By (7.2), $L(\xi, h) = Q_\xi(H^*; 0 < S < 1)$. Using (9.8) with $S = T$ and the fact that $H^* \in \mathcal{I}$, one has

$$Q_\xi(H^*; 0 < S < 1) = Q_\xi(\varphi(S)H^*)$$

provided $\varphi \in p\mathcal{B}$ with $\ell(\varphi) = 1$. Consequently $L(\xi, h) = Q_\xi(\varphi(S)H^*)$. Suppose $\psi(t) = t$. Then because of (8.13) and (8.21),

$$L(\xi, U_N 1) = \nu_N^\xi(1) = Q_\xi \int \varphi(t) N^*(dt)\,.$$

Combining these remarks establishes (10.7). ∎

If $F \in p\mathcal{F}^*$ and N is a HRM of X, then $M(dt) := F \circ \theta_t N(dt)$ defines a HRM of X. Moreover $M^*(dt) = F \circ \theta_t N^*(dt)$. Recall N^* and hence M^* are carried by $]\alpha, \beta]$. Applying (10.8) to M we obtain

$$\begin{aligned} (10.9) \qquad & L\left(\xi, P^\cdot \int_0^\infty F \circ \theta_t N(dt)\right) \\ & = Q_\xi \int_{-\infty}^\infty \varphi(t) F \circ \theta_t N^*(dt) \\ & = Q_\xi \left[\varphi(S) \int_{-\infty}^\infty F \circ \theta_t N^*(dt)\right], \end{aligned}$$

for $F \in p\mathcal{F}^*$ and $\varphi \in p\mathcal{B}$ with $\ell(\varphi) = 1$. If we take $\varphi = 1_{]0,1[}$ and combine this with (10.2), we obtain $\mathbb{P}_N^{\xi_k} \uparrow \mathbb{P}_N^\xi$ whenever $\xi_k \uparrow \xi \in \mathrm{Dis}$. Consequently using (10.5), it follows that

$$(10.10) \qquad \mu_n U \uparrow \xi \Longrightarrow P^{\mu_n} \int_0^\infty F \circ \theta_t N(dt) \uparrow \mathbb{P}_N^\xi(F)$$

provided $F \in p\mathcal{F}^*$ and N is a HRM of X.

When $\psi(t) = t^2$, (10.7) becomes

$$(10.11) \qquad L(\xi, P^\cdot[N(\mathbb{R}^+)^2]) = Q_\xi(N^*(\mathbb{R})^2; 0 < S < 1)\,.$$

A result that is known as Weil's energy formula.

We are next going to discuss capacity in the present context following [**GSt87**] and [**F88a**]. Actually one can develop these ideas in the framework of §9, but we will restrict our discussion to Markov processes. In what follows $m \in \mathrm{Dis}$ is fixed. We define the *capacity*, $\Gamma(B)$, of a set $B \in \mathcal{E}^e$ by

$$\Gamma(B) := L(m, P_B 1) = L(R_B m, 1)\,, \tag{10.12}$$

where the equality is just (4.12). When we want to indicate the dependence on m we shall write $\Gamma_m(B)$ in place of $\Gamma(B)$. We also fix once and for all a ST, S, such that $\alpha < S < \beta$ a.s. Q_m. We remind the reader that all named sets are in $\mathcal{E}^e$ unless stated otherwise. As usual T_B denotes the hitting time of B; however, see the word of caution following (7.6). We let τ_B be the extension of T_B to W defined in (7.7) Then τ_B is a stationary terminal time (STT) as defined in (6.22) and (7.8iii). As remarked in §7, $\tau_B = \inf\{t\colon Y_t \in B\}$ a.s. Q_ξ for each $\xi \in \mathrm{Exc}$ and also a.s. Q_ν for each entrance rule ν. Of course, when $B \in \mathcal{E}$ this holds identically. *In the sequel a property $p(w)$ defined for $w \in W$ will be said to hold* a.s. *provided it holds* a.s. Q_ν *for every entrance rule* ν; in particular, a.s. Q_ξ for every $\xi \in \mathrm{Exc}$. We shall often omit the "a.s." in those places where it is obviously required.

If $H := 1_{\{T_B < \infty\}}$, then $P_B 1 = P^{\cdot}(H)$, and H^* as defined in (7.1) is given by $H^* = 1_{\{\tau_B < \infty\}}$. Define

$$\rho(B) = \{t \colon Y_t \in B\}\,. \tag{10.13}$$

Note that $\{\tau_B < \infty\} = \{\rho(B) \neq \emptyset\}$. Consequently we obtain from (7.2) the following expression for $\Gamma(B)$ which is due to Fitzsimmons [**F88a**],

$$\begin{aligned} \Gamma(B) &= L(m, P_B 1) = Q_m(\tau_B < \infty;\, 0 < S < 1) \\ &= Q_m(\rho(B) \neq \emptyset;\, 0 < S < 1)\,. \end{aligned} \tag{10.14}$$

Next define $L_B := (\sup\{t \colon T_B \circ \theta_t < \infty\}) \vee 0$ and $\lambda_B := \sup\{t > \alpha \colon \tau_B \circ \theta_t < \infty\}$ where the supremum of the empty set is minus infinity. It follows from (7.6) that $L_B \in \mathcal{F}^*$ and because of (7.8) it is easy to check that λ_B is a ST. Clearly $L_B = (\sup\{t\colon X_t \in B\}) \vee 0$ a.s. and $\lambda_B = \sup\{t\colon Y_t \in B\}$ a.s. and these equalities hold identically when $B \in \mathcal{E}$. We call L_B (resp. λ_B) the *last exit time from* B *by* X (resp. Y). Since $\tau_B([\Delta]) = \infty$, $\lambda_B \leq \beta$ and clearly $\lambda_B > \alpha$ if $\lambda_B > -\infty$. Note that $\{L_B > 0\} = \{T_B < \infty\}$ and $\{\rho(B) \neq \emptyset\} = \{\lambda_B > -\infty\}$. Therefore

$$\Gamma(B) = Q_m(\lambda_B > -\infty;\, 0 < S < 1). \tag{10.15}$$

Since both τ_B and λ_B are stationary times

$$Q_m(\tau_B \in dt) = \hat{C}(B)\,dt;\; Q_m(\lambda_B \in dt) = C(B)\,dt \tag{10.16}$$

where

$$\hat{C}(B) = Q_m(0 < \tau_B < 1);\; C(B) = Q_m(0 < \lambda_B < 1). \tag{10.17}$$

It is understood that in (10.16) we are talking about measures on $\mathbb{R}$ and *not* on $[-\infty, \infty]$. Of course, both $C(B)$ and $\hat{C}(B)$ may take the value plus infinity. In an early paper **[GSt86]**, $C(B)$ and $\hat{C}(B)$ were called the capacity and cocapacity of B respectively. However, we shall not use this terminology. In this monograph the capacity of B is the number $\Gamma(B)$ defined in (10.12). Making use of (10.14) and (9.8) and the property of S one obtains

$$\begin{aligned} \Gamma(B) &= Q_m(\tau_B = -\infty;\, 0 < S < 1) \\ &\qquad + Q_m(\tau_B \in \mathbb{R};\, 0 < S < 1) \\ &= Q_m(\tau_B = -\infty;\, 0 < S < 1) + \hat{C}(B), \end{aligned} \tag{10.18}$$

and similarly

$$\Gamma(B) = Q_m(\lambda_B = \infty;\, 0 < S < 1) + C(B). \tag{10.19}$$

A set B is *transient* (resp. *cotransient*) provided $Q_m(\lambda_B = \infty) = 0$ (resp. $Q_m(\tau_B = -\infty) = 0$); if we want to emphasize m we shall say that B is *m-transient* (resp. *m-cotransient*). It is immediate that $C(B) = \Gamma(B)$ if B is transient and $\hat{C}(B) = \Gamma(B)$ if B is cotransient. Of course, $C(B) = \hat{C}(B) = \Gamma(B)$ if B is both transient and cotransient. In this case τ_B and λ_B have the same distribution under Q_m. More generally it follows from (9.25) that if B is transient and cotransient, then

$$(10.20) \qquad \Gamma(B) = Q_m(0 < a\tau_B + b\lambda_B < 1;\ \tau_B \in \mathbb{R};\ \lambda_B \in \mathbb{R})$$

provided $a, b \in \mathbb{R}$ with $|a + b| = 1$.

We now develop some of the properties of Γ. The proofs are taken from **[F88a]**. First note that it is immediate from (10.14) that $\Gamma(B) = 0$ if and only if B is *m-polar*; that is, $P^m(T_B < \infty) = 0$. Of course, this is equivalent to $\rho(B)$ is empty a.s. Q_m.

(10.21) Proposition. (i) $A \subset B \Longrightarrow \Gamma(A) \le \Gamma(B)$.
(ii) $B_n \uparrow B \Longrightarrow \Gamma(B_n) \uparrow \Gamma(B)$.
(iii) $\Gamma(A \cup B) + \Gamma(A \cap B) \le \Gamma(A) + \Gamma(B)$.

Proof. Points (i) and (ii) are clear in view of (10.14) since $A \subset B$ implies $\rho(A) \subset \rho(B)$ and $B_n \uparrow B$ implies $\rho(B_n) \uparrow \rho(B)$. For point (iii), note that if $f\colon [-\infty, \infty] \to [0, 1]$, then

$$f(\tau_A \wedge \tau_B) + f(\tau_A \vee \tau_B) = f(\tau_A) + f(\tau_B).$$

Let $f = 1_{[-\infty,\infty[}$ and integrate this equality with respect to Q_m over the set $\{0 < S < 1\}$ to obtain from (10.14)

$$(10.22) \qquad Q_m(\tau_A \vee \tau_B < \infty;\ 0 < S < 1) + \Gamma(A \cup B) = \Gamma(A) + \Gamma(B)$$

since $\tau_A \wedge \tau_B = \tau_{A\cup B}$. But $\tau_A \vee \tau_B \le \tau_{A\cap B}$ and, hence, (10.21iii) holds. ∎

Remarks. The property (10.21iii) is called the *strong subadditivity* of Γ. The identity (10.22) shows precisely when there is equality in (10.21iii).

Actually much more is valid. Given $B_1, \ldots, B_n$ let $\tau_i = \tau_{B_i}$, $1 \leq i \leq n$. A simple induction argument (just the familiar inclusion-exclusion relationship) shows that

$$(10.23) \quad \text{(i)} \;\; f(\tau_1 \wedge \cdots \wedge \tau_n) + \sum (-1)^k f(\tau_{i_1} \vee \cdots \vee \tau_{i_k}) = 0,$$

$$\text{(ii)} \;\; f(\tau_1 \vee \cdots \vee \tau_n) + \sum (-1)^k f(\tau_{i_1} \wedge \cdots \wedge \tau_{i_k}) = 0,$$

for any real valued function f on $[-\infty, \infty]$ where the sums are over all non-empty finite subsets $\{i_1, \ldots, i_k\}$ of $\{1, 2, \ldots, n\}$. Again taking expectations relative to Q_m over $\{0 < S < 1\}$ in (10.23ii) we obtain

$$Q_m(\tau_{B_1} \vee \cdots \vee \tau_{B_n} < \infty; 0 < S < 1) = \sum (-1)^{k+1} \Gamma(B_{i_1} \cup \cdots \cup B_{i_k})$$

and hence

$$(10.24) \quad \Gamma(B_1 \cap \cdots \cap B_n) \leq \sum (-1)^{k+1} \Gamma(B_{i_1} \cup \cdots \cup B_{i_k}).$$

Since $\Gamma(B)$ may be infinite, the precise statement of these formulas should be written with the terms involving even k moved to the other side. Then all terms are positive and no factors of minus one appear. In particular when $n = 2$, (10.24) reduces to (10.21iii). The set function Γ is *alternating of order* n provided, given A and $B_1, \ldots, B_n$, one has

$$\Gamma(A) + \sum (-1)^k \Gamma(A \cup A_{i_1} \cup \cdots \cup A_{i_k}) \leq 0,$$

where again the sum is over all nonempty subsets of $\{1,2,\ldots,n\}$ and the sum must be rearranged by taking the negative terms to the other side of the inequality. Γ is *alternating of order infinity* provided it is alternating of order n for each $n \geq 1$. It is well known and easily checked that an increasing set function is alternating of order n if and only if it satisfies (10.24). Consequently Γ is alternating of order infinity. Using (10.23ii) with $f = 1_{]0,1[}$ we obtain in light (10.17)

$$Q_m(0 < \tau_{B_1} \vee \cdots \vee \tau_{B_n} < 1) = \sum (-1)^k \hat{C}(B_{i_1} \cup \cdots \cup B_{i_k}) .$$

If at least one B_i is cotransient the term on the left side of this last display is equal to

$$\begin{aligned} Q_m(\tau_{B_1} \vee \cdots \vee \tau_{B_n} < \infty;\ 0 < S < 1) &\geq \Gamma(B_1 \cap \cdots \cap B_n) \\ &= \hat{C}(B_1 \cap \cdots \cap B_n) . \end{aligned}$$

Hence $\hat{C}$ satisfies (10.24) provided at least one B_i is cotransient. Similarly starting from (10.23i) with the τ_i replaced by λ_i, $1 \leq i \leq n$, one sees that C satisfies (10.24) provided at least one B_i is transient.

Before proceeding we shall show that $\Gamma(B)$ agrees with the usual notion of capacity in a familiar situation. Suppose that X and $\hat{X}$ are standard processes in strong duality relative to m as described above (4.15). Suppose that there exists a measure π_B on E such that

$$\phi_B(x) := P_B 1(x) = \int u(x,y) \pi_B(dy) = \langle \pi_B, u(x, \cdot) \rangle .$$

Then π_B is called the *capacitary measure of* B and $C(B) := \pi_B(1)$ the capacity of B. See [**BG**, VI-§4] or, in a more general context, [**G84**]. Since $m \in$ Dis there exist potentials $\mu_n U \uparrow m$. But then the coexcessive functions $h_n(y) := \langle \mu_n, u(\cdot, y) \rangle$ increase to 1 a.e. m and hence everywhere since m is a reference measure. Therefore

$$\Gamma(B) = L(m, P_B 1) = \lim_n \mu_n(P_B 1) = \lim_n \langle \pi_B, h_n \rangle = \pi_B(1) .$$

On the other hand suppose there exists a measure $\hat{\pi}_B$ on E such that

$$\hat{\phi}_B(x) := \hat{P}_B 1(x) = \int \hat{\pi}_B(dy) u(y,x) = \langle \hat{\pi}_B, u(\cdot, x) \rangle .$$

Then $\hat{\pi}_B$ is called the *cocapacitary measure of* B and $\hat{C}(B) := \hat{\pi}_B(1)$ the cocapacity of B. By (4.15), $R_B m = \hat{\phi}_B m = \hat{\pi}_B U$, and so

$$\Gamma(B) = L(m, P_B 1) = L(R_B m, 1) = \hat{\pi}_B(1).$$

Thus if π_B (resp. $\hat{\pi}_B$) exists, then $\Gamma(B) = C(B)$ (resp. $\Gamma(B) = \hat{C}(B)$), and if both π_B and $\hat{\pi}_B$ exist, then $C(B) = \Gamma(B) = \hat{C}(B)$.

In order to complete this comparison we now show that $C(B)$ and $\hat{C}(B)$ as defined in the preceding paragraph agree with the definitions in (10.17). This will be an immediate consequence of the above discussion and (10.18) and (10.19) once we show that the existence of π_B (resp. $\hat{\pi}_B$) implies that B is transient (resp. cotransient). We saw at the end of the previous paragraph that the existence of $\hat{\pi}_B$ implies that $R_B m = \hat{\pi}_B U \in \mathrm{Pot}$. Suppose π_B exists. If $f \geq 0$ with $m(f) < \infty$, then

$$m(f P_t \phi_B) = \langle \pi_B, \hat{P}_t \hat{U} f \rangle \longrightarrow 0$$

as $t \to \infty$ because $\langle \pi_B, \hat{U} f \rangle \leq m(f) < \infty$. Therefore $P_t \phi_B \to 0$ a.e. m as $t \to \infty$. The following proposition valid for an arbitrary right process satisfying (6.2) now completes the identification of the two expressions for $C(B)$ and $\hat{C}(B)$.

(10.25) Proposition. *Let* $B \in \mathcal{E}^e$. *Then* (i) B *is transient if and only if* $P_t \phi_B \to 0$ *a.e.* m *as* $t \to \infty$, *and* (ii) B *is cotransient if and only if* $R_B m \in \mathrm{Pur}$.

Proof. Since $\lambda_B = t + L_B \circ \theta_t$ if $\alpha < t < \lambda_B$ one has

$$Q_m(\alpha < t;\ \lambda_B = \infty) = Q_m[P^{Y(t)}(L_B = \infty)] = P^m(L_B = \infty).$$

But $\{\lambda_B = \infty\} = \bigcup_{r \in \mathbb{Q}} \{\alpha < r;\ \lambda_B = \infty\}$ and so B is transient if and only if $P^m(L_B = \infty) = 0$. Now

$$P_t \phi_B(x) = P^x(t + T_B \circ \theta_t < \infty) = P^x(L_B > t).$$

Therefore $P_t\phi_B(x) \to P^x(L_B = \infty)$ as $t \to \infty$ and combining these observations establishes (i). For (ii), given $f \in p\mathcal{E}$ with $m(f) < \infty$, one has $Q_m(f \circ Y_r) = m(f) < \infty$ for each $r \in \mathbb{Q}$. Since $\alpha \le \tau_B$, this justifies the following computation:

$$\begin{aligned} Q_m[\tau_B = -\infty;\ f \circ Y_r] &= Q_m[\tau_B = -\infty;\ f \circ Y_0] \\ &= \lim_{t\to\infty} Q_m[\tau_B < -t;\ f \circ Y_0] = \lim_{t\to\infty} Q_m[\tau_B < 0;\ f \circ Y_t] \\ &= \lim_{t\to\infty} Q_m[\tau_B < 0;\ P_t f \circ Y_0] = \lim_{t\to\infty} R_B m(P_t f), \end{aligned}$$

where the last equality comes from (7.3) and (7.9). ∎

We noted above that when X and $\hat{X}$ are standard processes in strong duality relative to m and a cocapacitary measure $\hat{\pi}_B$ exists for B, then $R_B m = \hat{\pi}_B U$. This motivates the following definition. Once again X is a fixed right process satisfying (6.2) and $m \in \text{Dis}$ is fixed.

(10.26) Definition. A set $B \in \mathcal{E}^e$ is an *equilibrium set* provided $R_B m \in \text{Pot}$. In this case the unique σ-finite measure γ_B satisfying $R_B m = \gamma_B U$ is called the *capacitary measure* of B.

(10.27) Remarks. Note that in the duality situation discussed above $\gamma_B = \hat{\pi}_B$ which is called the cocapacitary measure under duality hypotheses. This is an unfortunate historical accident that goes back to Hunt's original memoir and is perpetuated in [**BG**] and the literature of Markov processes. However, Port and Stone [**PS71**] in their development of the potential theory for infinitely divisible processes did indeed interchange the terminology for capacity and cocapacity. Under the present hypotheses I am convinced that γ_B defined in (10.26) should be called the capacitary measure and I have decided to do so. But in an attempt to minimize the confusion I have adopted the notation γ_B rather than π_B. If we need to emphasize the dependence on m we shall say that B is an m-equilibrium set and write γ_B^m in place of γ_B and call γ_B^m the m-capacitary measure of B.

Because of (5.23) any subset (in $\mathcal{E}^e$) of an equilibrium set is an equilibrium set. If B is an equilibrium set, then $\Gamma(B) = L(R_B m, 1) = \gamma_B(1)$ and (10.25ii) implies that B is cotransient. The next proposition (especially in conjunction with the Remark (10.29) following its proof) supplies a ready source of equilibrium sets. For its statement recall from §4 that σ_B is the Lebesgue penetration time of B.

(10.28) Proposition. *Let $B \in \mathcal{E}^e$ be such that $\sigma_B = T_B$ a.s. P^m; for example, B finely open. Suppose there exists $\pi = \mu U \in \text{Pot}$ with $m \le \pi$ on B. Then B is an equilibrium set and $\Gamma(B) \le \mu(1)$.*

Proof. Because of (4.26), $R_B m = R^*_B m \le \pi$ and so $R_B m \in \text{Pot}$. Then $\Gamma(B) = L(R_B m, 1) \le L(\pi, 1) = \mu(1)$. ∎

(10.29) Remark. In [**F90**, §2], Fitzsimmons has shown that if $\xi, m \in \text{Exc}$ with $\xi << m$, then there exists a version h of $d\xi/dm$ that is finely continuous off an m-polar Borel set D such that $E \backslash D$ is absorbing for X, at least when X is a Borel right process. Given $\mu U \in \text{Pot}$ let $\eta = m + \mu U$. Then there exist versions g and h of $dm/d\eta$ and $d(\mu U)/d\eta$ respectively which are finely continuous off a Borel m-polar set D and $E \backslash D$ is finely open. Thus for any $a > 0$, $B := \{g < ah\}$ satisfies the hypotheses of (10.28) and so B is equilibrium set with $\Gamma(B) \le a\mu(1)$.

It is immediate from (7.5iii) that if B is an equilibrium set, then

$$(10.30) \qquad \gamma_B(f) = Q_m[f(Y_{\tau_B+});\ 0 < \tau_B < 1;\ W(h) \circ b_{\tau_B}]$$

where $W(h)$ is defined following the proof of (6.11). Note that $b_s \sigma_t = \sigma_t b_{s+t}$ for $s, t \in \mathbb{R}$. As a result $b_{\tau_B} \circ \sigma_t = \sigma_t b_{\tau_B}$ on $\{\tau_B \in \mathbb{R}\}$. Since $W(h)$ is invariant it now follows that $1_{\mathbb{R}}(\tau_B) W(h) \circ b_{\tau_B}$ is invariant. Recall that S is a fixed ST with $Q_m(S \notin \mathbb{R}) = 0$. Consequently by (6.27) or (9.8)

$$(10.31)\ \gamma_B(f) = Q_m[f(Y_{\tau_B+});\ 0 < S < 1;\ W(h) \circ b_{\tau_B};\ \tau_B < \infty].$$

A set $B \in \mathcal{E}^e$ is *strongly cotransient* provided $Q_m(\tau_B = \alpha) = 0$. (Recall $\tau_B \geq \alpha$.) Since $(W(h) \circ b_t) \cap \{\alpha < t < \beta\} = \{\alpha < t < \beta\}$, it follows from (7.5i) and (6.19) that a strongly cotransient set B is an equilibrium set and that

$$(10.32) \quad \gamma_B(f) = Q_m[f \circ Y_{\tau_B};\ 0 < \tau_B < 1] = Q_m[f \circ Y_{\tau_B};\ 0 < S < 1;\ \tau_B < \infty].$$

Perhaps the nicest way to look at (10.30) is as follows. Recall the definition of the process (Y_t^*) in (6.12). Define

$$(10.33) \qquad \tau_B^* := \inf\{t : Y_t^* \in B\}.$$

It is readily checked that τ_B^* is a SST and that $\alpha \leq \tau_B^* \leq \tau_B$. Since $Y_t^* = Y_t$ if $t > \alpha$, $\tau_B^* = \tau_B$ whenever $\tau_B^* > \alpha$. Moreover it is easily verified that

$$(W(h) \circ b_t) \cap \{\alpha = t\} = \{\alpha = t;\ Y_t^* \in E\}.$$

As a result one obtains from (10.30) or (10.31)

$$(10.34) \quad \gamma_B(f) = Q_m[f \circ Y^*(\tau_B^*);\ 0 < \tau_B^* < 1] = Q_m[f \circ Y^*(\tau_B^*);\ 0 < S < 1;\ \tau_B^* < \infty].$$

If we write P_B^* for the Palm measure, $P_{\tau_B^*}^m$, of τ_B^* relative to Q_m, then (10.34) states that γ_B is the restriction of $Y_0^*(P_B^*)$ to E. Of course, by (7.5iii) for any $B \in \mathcal{E}^e$ the potential part of $R_B m$ is given by $\gamma_B U$ where γ_B is defined in (10.30). However following the usual convention, we call γ_B the capacitary measure of B only when B is an equilibrium set.

(10.35) Remarks. Suppose X and $\hat{X}$ are standard process in strong duality relative to m and B is strongly transient in the sense that $P^m(L_B \geq \zeta) = 0$. Then one can show that the "capacitary" measure π_B, of B as defined in [**BG**], for example, exists and is given by

$$\pi_B(f) = Q_m[f(Y_{\lambda_B -});\ 0 < \lambda_B < 1].$$

One may investigate this measure for any right process X such that $t \to X_t$ has left limits in E on $]0,\zeta[$. See, for example, [**GSt86**] where several other variants are described also.

It is clear from (10.30) or (10.34) that γ_B is carried by the closure of B for any B. For an equilibrium set we can say more.

(10.36) Theorem. *Let B be an equilibrium set. Then γ_B is carried by the fine closure of B.*

Proof. Let $m = \mu U + \eta$ be the decomposition of m into its potential and harmonic parts. Then $R_B m = \mu P_B U + R_B \eta$. Hence $R_B\eta \le R_B m \in \text{Pot}$, $R_B\eta = \nu U$, and $\gamma_B = \mu P_B + \nu$. But $P_B(x,\cdot)$ is carried by the fine closure, $\overline{B}^f$, of B for each x and, hence, so is μP_B. Thus it suffices to prove that ν is carried by $\overline{B}^f$. In other words it suffices to prove (10.36) under the additional assumption that $m \in \text{Har}$.

We first claim that under these hypotheses ($m \in \text{Har}$ and $R_B m \in \text{Pot}$), $Q_m(\alpha = \tau_B) = 0$. To this end first observe that because of (6.19) and (7.5i)

$$0 = Q_{R_B m}[W(h)^c] = Q_m[W(h)^c \circ b_{\tau_B};\ \tau_B < \beta]$$

which implies $Q_m[\alpha = \tau_B;\ W(h)^c \circ b_{\tau_B}] = 0$. Since $W(h)^c \circ b_\alpha = W(h)^c$ and Q_m is carried by $W(h)^c$, $Q_m(\alpha = \tau_B) = 0$ as claimed. Therefore B is strongly cotransient. Let g be the indicator of the complement of $\overline{B}^f$. Then from (10.32)

$$\begin{aligned}\gamma_B(g) &= Q_m[g \circ Y_{\tau_B};\ \alpha < \tau_B;\ 0 < \tau_B < 1]\\ &\le \sum Q_m[g \circ Y_{\tau_B};\ \alpha < r < \tau_B < 1],\end{aligned}$$

where the sum is over all rationals r with $0 < r < 1$. But $Y_{\tau_B} = X_{T_B} \circ \theta_r$ if $\alpha < r < \tau_B$, and so

$$\begin{aligned}&Q_m[g \circ Y_{\tau_B};\ \alpha < r < \tau_B < 1]\\ &\qquad \le Q_m[P^{Y(r)}(g \circ X_{T_B});\ \alpha < r] = 0,\end{aligned}$$

since $X_{T_B} \in \overline{B}^f$ a.s. on $\{T_B < \infty\}$. ∎

We turn now to a brief discussion of q-capacities; that is, capacity relative to X^q —the q-subprocess of X. For this we fix $m \in \text{Exc}$. *Note we no longer assume* $m \in \text{Dis}$. For $B \in \mathcal{E}^e$ and $q \geq 0$, the q-capacity of B, $\Gamma^q(B) = \Gamma^q_m(B)$ is defined by

$$(10.37) \qquad \Gamma^q(B) = \Gamma^q_m(B) := L^q(m, P^q_B 1) = L^q(R^q_B m, 1).$$

Of course, the last equality in (10.37) is just (4.12). If $q = 0$, $\Gamma^0_m(B) = \Gamma_{m_d}(B)$ as defined in (10.12). From now on we shall write $\Gamma_m(B) = \Gamma^0_m(B)$ for a general $m \in \text{Exc}$. Since $\Gamma_m(B) = \Gamma_{m_d}(B)$ our notation is consistent. Because $m \in \text{Dis}^q$ if $q > 0$, everything that we have proved for $\Gamma_m(B)$ under the assumption that $m \in \text{Dis}$ extends to $\Gamma^q_m(B)$ for $q > 0$ and $m \in \text{Exc}$. We shall use these results for Γ^q, $q > 0$ without special mention. The next result details the relationship between Γ^q and Γ^r. It is an immediate consequence of (4.37).

(10.38) Theorem. *Let* $0 \leq r < q$ *and* $B \in \mathcal{E}^e$. *Then*

$$\begin{aligned} \Gamma^q(B) &= \Gamma^r(B) + (q-r) R^r_B m(P^q_B 1) \\ &= \Gamma^r(B) + (q-r) R^q_B m(P^r_B 1). \end{aligned}$$

If $\Gamma^q(B) < \infty$ for one value of $q > 0$, then $q \to \Gamma^q(B)$ is a subordinator exponent; that is, it has a completely monotone derivative. For the proof we need several lemmas. These are special cases of results in [**FG88**] and are also closely related to [**S**, (75.5)]. For their statement we fix an exact terminal time T and let (Q_t) and (V^q) be the semigroup and resolvent of X killed at T; that is

$$Q_t f(x) = P^x[f \circ X_t;\ t < T];\ V^q f(x) = P^x \int_0^T e^{-qt} f \circ X_t \, dt.$$

Let η be an s-finite measure on E satisfying $\eta Q_t \leq \eta$ for each $t \geq 0$. Note that any $\eta \in \text{Exc}(X)$ satisfies this condition.

(10.39) Lemma. *Define* $\varphi(t) := P^\eta(0 < T \leq t)$. *If* $\varphi(t)$ *is finite for one* $t > 0$, *then* φ *is a finite, increasing, continuous, concave function on* $[0, \infty[$ *with* $\varphi(0) = 0$.

Proof. Obviously φ is increasing. If $a_t(x) := P^x(0 < T \leq t)$ so that $\varphi(t) = \eta(a_t)$, then

$$(10.40) \qquad \varphi(t+s) = \varphi(t) + \eta Q_t(a_s) \leq \varphi(t) + \varphi(s).$$

Suppose $\varphi(t) < \infty$ for one $t > 0$. It now follows that $\varphi(t) < \infty$ for all $t \geq 0$. Consequently, φ is right continuous on $[0, \infty[$. For $t > 0$ define

$$\Delta\varphi(t) := \varphi(t) - \varphi(t-) = P^\eta(T = t),$$
$$\Delta a_t(x) := a_t(x) - a_{t-}(x) = P^x(T = t).$$

If $0 < s < t$, one then has

$$\Delta\varphi(t) = P^\eta(s < T = t) = \eta Q_s(\Delta a_{t-s}) \leq \Delta\varphi(t-s).$$

Therefore $t \to \Delta\varphi(t)$ is decreasing and since $\Delta\varphi(t) > 0$ for at most countably many t, we see that $\Delta\varphi = 0$ and φ is continuous. Clearly $t \to \eta Q_t$ is decreasing. Hence (10.40) implies that for $0 < s < t$ and $h > 0$

$$\varphi(t+h) - \varphi(t) \leq \varphi(s+h) - \varphi(s).$$

Setting $t = s + h$ this becomes

$$\frac{1}{2}\left[\varphi(s+2h) + \varphi(s)\right] \leq \varphi(s+h) = \varphi\left[\frac{1}{2}(s+2h) + \frac{1}{2}s\right].$$

Therefore φ is "midpoint concave"; hence concave being continuous. ∎

Here are some consequences of (10.39). Suppose $\varphi(t) < \infty$ for some $t > 0$. Then the right hand derivative, φ^+, of φ exists for all $t \geq 0$ and is finite for $t > 0$. It is possible that $\varphi^+(0) = \infty$. The function φ^+ is decreasing and right continuous on $[0, \infty[$ and $\varphi^+(\infty) := \lim_{t\to\infty} \varphi^+(t) < \infty$. Being concave, φ is absolutely continuous and so ($\varphi(0) = 0$)

$$\varphi(t) = \int_0^t \varphi^+(s)\, ds,\ t \geq 0. \tag{10.41}$$

It follows from (10.41) that $\varphi^+(\infty) = \lim_{t\to\infty} \varphi(t)/t$.

(10.42) Lemma. *Let φ be as in (10.39). Define $\psi(q) := P^\eta(e^{-qT}; T > 0)$ for $q > 0$. Suppose $\psi(q_0) < \infty$ for some $q_0 > 0$. Then for each $q > 0$, $\psi(q) = \int_0^\infty e^{-qt}\, d\varphi(t) < \infty$ and*

$$q\psi(q) = \varphi^+(\infty) + \int_0^\infty (1 - e^{-qt})\nu(dt) \tag{10.43}$$

where $\nu(dt) = -1_{]0,\infty[}(t)\, d\varphi^+(t)$.

Proof. The hypothesis $\psi(q_0) < \infty$ implies that $\varphi(t) < \infty$ for all $t \geq 0$. Hence for each $q > 0$, $\psi(q) = \int_0^\infty e^{-qt}\, d\varphi(t)$. Consequently using the properties of φ we find for $q > 0$

$$\begin{aligned} q\psi(q) &= q \int_0^\infty e^{-qt}\, d\varphi(t) = q \int_0^\infty e^{-qt} \varphi^+(t)\, dt \\ &= (1 - e^{-qt})\varphi^+(t)\Big|_0^\infty - \int_0^\infty (1 - e^{-qt})\, d\varphi^+(t). \end{aligned}$$

But φ^+ is decreasing and so $t\varphi^+(t) \leq \int_0^t \varphi^+(s)\, ds = \varphi(t)$. Hence $t\varphi^+(t) \to 0$ as $t \to 0$ and the preceding formula becomes (10.43). Using $\psi(q_0) < \infty$ once again we see that $\int_0^\infty (t \wedge 1)\nu(dt) < \infty$ and consequently $\psi(q) < \infty$ for all $q > 0$. ∎

Since the hitting time of a set $B \in \mathcal{E}^e$ and the hitting time of its fine closure $\overline{B}^f$ are almost surely equal it is obvious from the definition that $\Gamma^q(B) = \Gamma^q(\overline{B}^f)$ for all $q \geq 0$. Thus in discussing $\Gamma^q(B)$ there is no loss of generality in supposing that B is finely closed.

(10.44) Theorem. *Let $m \in \mathrm{Exc}$ and $B \in \mathcal{E}^e$ be finely closed. Suppose $\Gamma^q(B) < \infty$ for some $q > 0$. Then*

$$\Gamma^q(B) = \Gamma(B) + qm(B) + \int_0^\infty (1 - e^{-qt})\nu_B(dt)$$

where ν_B is a measure on $]0, \infty[$ such that $\int_0^\infty (t \wedge 1)\nu_B(dt) < \infty$. In particular $\Gamma(B) = \downarrow \lim_{q \downarrow 0} \Gamma^q(B)$.

Proof. From (10.38), $\Gamma^q(B) = \Gamma(B) + qR_Bm(P_B^q 1)$. Now

$$\begin{aligned} R_Bm(P_B^q 1) &= P^{R_B m}(e^{-qT_B}) \\ &= R_Bm(B^r) + P^{R_B m}(e^{-qT_B};\ T_B > 0)\,. \end{aligned}$$

Since $R_Bm = m$ on B, $B^r \subset B$, and m, being excessive, vanishes on the semipolar set $B \backslash B^r$, $R_Bm(B^r) = m(B)$. Next apply (10.42) with $\eta = R_Bm$ and $T = T_B$ to obtain

$$qP^{R_B m}(e^{-qT_B};\ T_B > 0) = c + \int_0^\infty (1 - e^{-qt})\nu_B(dt)\,,$$

where ν_B is a measure as in the statement of (10.44) and $c = \varphi^+(\infty)$ where $\varphi(t) := P^{R_B m}(0 < T_B \leq t)$. Combining these observations will yield (10.44) provided we show $c = 0$.

To this end we first claim that $(R_Bm)Q_t \to 0$ as $t \to \infty$ where (Q_t) is the semigroup of X killed at T_B. To see this let $f \geq 0$ with $m(f) < \infty$. Then

$$\begin{aligned} R_Bm(Q_tf) &= Q_m[Q_tf(Y_0);\ \tau_B < 0] \\ &= Q_m[f \circ Y_t;\ t < T_B \circ \theta_0;\ \tau_B < 0;\ \alpha < 0] \\ &= Q_m[f \circ Y_0;\ t < T_B \circ \theta_{-t};\ \tau_B < -t,\ \alpha < -t] \end{aligned}$$

where the third equality uses the invariance of Q_m and $\theta_0\sigma_{-t} = \theta_{-t}$. Split this last integral into an integral, I_1, over $\{\alpha > -\infty\}$ and an integral I_2 over $\{\alpha = -\infty\}$. Then

$$I_1 \leq Q_m[f \circ Y_0;\ -\infty < \alpha < -t] \longrightarrow 0$$

as $t \to \infty$, and

$$I_2 = Q_m[f \circ Y_0;\ \alpha = -\infty;\ \tau_B < -t,\ -t + T_B \circ \theta_{-t} > 0] .$$

But $-t + T_B \circ \theta_{-t}$ decreases to τ_B as $t \to \infty$ on $\{\alpha = -\infty\}$, and it follows that $I_2 \to 0$ as $t \to \infty$. Hence $\lim_{t \to 0} (R_B m)Q_t = 0$. If $a_t(x) = P^x(0 < T_B \leq t)$, then $\infty > \varphi(1) = R_B m(a_1) \geq (R_B m)Q_t(a_1) \to 0$ as $t \to \infty$. Consequently (10.40) with η replaced by $R_B m$ forces $\varphi(t+1) - \varphi(t) \to 0$ as $t \to \infty$. But φ^+ is decreasing and so

$$\varphi^+(t+1) \leq \int_t^{t+1} \varphi^+(s)\, ds = \varphi(t+1) - \varphi(t) \longrightarrow 0 .$$

Therefore $c = \varphi^+(\infty) = 0$. ■

Remarks. It follows from (10.44) that if $\Gamma^q(B) < \infty$ for some $q > 0$, then $m(B) < \infty$. The measure ν_B in (10.44) may be further identified in terms of the exit system associated with B. See (11.18). Of course, ν_B is uniquely determined by $q \to \Gamma^q(B)$, and hence by B, under the hypothesis of (10.44). It is possible to have $\Gamma^q(B) = \infty$ for all $q > 0$ and $\Gamma(B) < \infty$. Consider the following example: X is translation to the right at unit speed on $\mathbb{R}$, m is Lebesgue measure, and $B =]-\infty, 0[$. Since $\epsilon_{-n}U \uparrow m$, $\Gamma(B) = 1$. But $R_B m = m$ while $P_B^q 1(x) = 1$ if $x < 0$ and $= 0$ if $x \geq 0$. Therefore $\Gamma^q(B) = \Gamma(B) + qR_B m(P_B^q 1) = \infty$.

We refer the reader to **[GSt87]** and especially **[GSt88]** for additional properties of the capacity function $\Gamma^q(B)$, $q \geq 0$. In these references a more general capacity is discussed in which

1 is replaced by an arbitrary $u \in S$ with $u < \infty$ a.e. m. The study of these more general capacities may be reduced to the case $u = 1$ by means of an h-transform with $h = u$. The interested reader should consult these references. See also **[G84]**.

11. Exit Systems and Applications

In this section we shall give several applications of the "exit system" associated with an exact terminal time. The first and most important of these is an unpublished asymptotic formula of Fitzsimmons that generalizes Spitzer's formula [**Sp64**]. The nicest treatment to date of a "Spitzer's formula" for (Borel) right processes is contained in [**St87**]. See also [**GSt86**] and [**G84**] for earlier versions. Exit systems are discussed in [**S**, §74] for a general right process X. However, there is an important simplification when X is a Borel right process. To keep the discussion as simple as possible we shall consider the case of a Borel right process first, and then indicate the extension to more general processes.

We assume that X is a Borel right process until further notice. Then X satisfies (6.2) and so by our convention, X is defined on the sample space, Ω, described below (6.5). We fix an exact terminal time T of X. Define $T_t = t + T \circ \theta_t$. Then $t \to T_t$ is increasing and right continuous on $[0, \infty[$ and constant on each interval of the form $[s, T_s[$, $s \geq 0$. We define two random sets

$$(11.1) \qquad M = M(\omega) := \{T_t(\omega);\ t > 0\}^- \cap\ ]0, \zeta(\omega)[$$

where the bar "$-$" denotes closure, and

$$(11.2) \qquad G = G(\omega) := \{t \in\ ]0, \zeta(\omega)[;\ T_{t-}(\omega) = t < T_t(\omega)\}.$$

Then M is a closed homogeneous (that is, $s + t \in M \iff s \in M(\theta_t)$ for $s > 0$, $t \geq 0$) random set and G is the set of left endpoints contained in $]0, \zeta[$ of the intervals contiguous to M. Of course, G is also a homogeneous random set. The next result is a special case of [**S**, 74.12]. See also [**Ma75**].

(11.3) Proposition. *There exists a proper kernel* $\hat{P}^x(d\omega)$ *from* $(E, \mathcal{E}^*)$ *to* $(\Omega, \mathcal{F}^*)$ *and an* AF, A, *of* X *with a bounded one potential such that:*

(i) $0 < \hat{P}^x(1 - e^{-T}) \leq 1, \ \hat{P}^x(T = 0) = 0$.

(ii) *If* $Z \geq 0$ *is optional and* $F \in p\mathcal{F}^*$, *then*

$$P^x \sum_{s \in G} Z_s F \circ \theta_s = P^x \int_0^\infty Z_s \hat{P}^{X(s)}(F)\, dA_s , \tag{11.4}$$

for each $x \in E$.

Remarks. $(\hat{P}^x, A)$ is called an *exit system for* T. The reason $\hat{P}$ is a kernel from $(E, \mathcal{E}^*)$ to $(\Omega, \mathcal{F}^*)$ rather than to $(\bar{\Omega}, \bar{\mathcal{F}}^*)$ as in [**S**, 74.12] is that we are assuming that X is a Borel right process and, hence, E is Lusinian. See the note above (74.12) on page 355 of [**S**]. Note also that with M defined as in (11.1), $T = \inf\{t > 0 : t \in M\}$ and so our notation agrees with that in [**S**].

We next are going to extend (11.4) to Y defined on W following [**FM86**]. Let τ be the extension of T to W defined in (7.7). Define $T_t = t + T \circ \theta_t$ on $\{\alpha < t\}$. We also extend M to W by

$$M^* = M^*(w) := \bigcup_{\alpha(w) < t < \beta(w)} (t + M(\theta_t w)) .$$

Since $t + M(\theta_t) \subset t +]0, \zeta(\theta_t)[=]t, \beta[$ when $\alpha < t < \beta$, it is clear that $M^* \subset]\alpha, \beta[$. Suppose $\alpha < s < t < u < \beta$. Then

$$u - t + (t - s) \in M(\theta_s) \Longleftrightarrow u - t \in M(\theta_t) ,$$

and so $(s + M(\theta_s)) \cap]t, \beta[= t + M(\theta_t)$. Hence $M^* \cap]t, \beta[= t + M(\theta_t)$. It now follows that M^* is closed in $]\alpha, \beta[$. Let $J_t = 1_M(t)$ and $J_t^* = 1_{M^*}(t)$ so J_t is defined on Ω and J_t^* on W. Let G^* be the set of left endpoints contained in $]\alpha, \beta[$ of the maximal open intervals in $[\alpha, \beta[\setminus M^*$. If $\alpha < \tau$, then

$]\alpha, \tau[$ is such an interval but $\alpha \notin G^*$. Clearly $\tau = \inf M^*$. The following lemma collects some elementary facts that we shall need.

(11.5) Lemma. *With the above notation:*

(i) $J^*_{s+t} = J^*_s \circ \sigma_t$ *for* $s, t \in \mathbb{R}$;

(ii) $J^*_t = J_{t-s} \circ \theta_s$ *for* $\alpha < s < t < \beta$;

(iii) $M^* = \{T_t \colon \alpha < t < \beta\}^- \cap]\alpha, \beta[$;

(iv) $t + T \circ \theta_t = \inf\{u > t \colon u \in M^*\}$ *if* $\alpha < t < \beta$;

(v) $G^* = \{t \in]\alpha, \beta[\colon T_{t-} = t < T_t\}$.

Proof. (i) is immediate from the definition of M^* and the fact that $\theta_t \sigma_s = \theta_{t+s}$ for $s, t \in \mathbb{R}$. For (ii) fix $\alpha < s < t < \beta$. Then using the observations following the definition of M^*

$$t \in M^* \Longleftrightarrow t \in s + M(\theta_s) \Longleftrightarrow t - s \in M(\theta_s).$$

For (iii) fix $\alpha < t < \beta$ and choose $\alpha < r < t$. Then $T_t = r + (t - r) + T(\theta_{t-r}\theta_r)$. But $t - r + T(\theta_{t-r}\omega) \in M(\omega)$ for $\omega \in \Omega$ and hence $T_t \in r + M(\theta_r) \subset M^*$. Therefore if N^* is the set on the right side of (11.5iii), then $N^* \subset M^*$. Conversely let $t \in M^*$. Then $\alpha < t < \beta$ and so we may choose r with $\alpha < r < t$. Hence

$$\begin{aligned} t &\in r + M(\theta_r) \\ &= r + \{u + T(\theta_u \theta_r);\ 0 < u < \zeta(\theta_r)\}^- \cap]0, \zeta(\theta_r)[\\ &= \{u + T(\theta_u);\ r < u < \beta\}^- \cap]r, \beta[\subset N^*. \end{aligned}$$

Next (iv) follows from (iii) and the fact that $\alpha < s < t < s + T \circ \theta_s$ implies $t + T \circ \theta_t = s + T \circ \theta_s$. Finally (v) follows from (iii) and (iv). We leave the details to the reader. ∎

(11.6) Proposition. *Let* $m \in \text{Exc}$ *and let* $(\hat{P}^x, A)$ *be an exit system for* T. *Then* $\nu := \nu^m_A$, *the Revuz measure of* A *relative to* m, *is* σ*-finite, and for* $H \in p(\mathcal{B} \otimes \mathcal{E}^* \otimes \mathcal{F}^*)$ *one has*

$$(11.7)\quad Q_m \sum_{s \in G^*} H(s, Y_s, \theta_s) = \int dt \int \nu(dx) \hat{P}^x[H(t, x, \cdot)].$$

Proof. Let $0 < h \le 1$ with $m(h) < \infty$. Then

$$U^1 h(x) \ge P^x \sum_{s\in G} \int_s^{T_s} e^{-t} h\circ X_t\,dt$$

$$= P^x \sum_{s\in G} e^{-s} \left(\int_0^T e^{-t} h\circ X_t\,dt\right)\circ\theta_s$$

$$= P^x \int_0^\infty e^{-s}\hat{P}^{X(s)}\left(\int_0^T e^{-t} h\circ X_t\,dt\right)\,dA_s$$

$$= U_A^1 g(x)\,,$$

where $g := \hat{P}^{\cdot} \int_0^T e^{-t} h(X_t)\,dt$. Because of (11.3i) and $0 < h \le 1$, we see that $0 < g \le 1$. By (8.10) and (8.13)

$$\nu_A^m(g) = L^1(m, U_A^1 g) \le L^1(m, U^1 h) = m(h) < \infty\,,$$

and so $\nu = \nu_A^m$ is σ-finite. Let $f \in p\mathcal{E}$ and $F \in p\mathcal{F}^0$. Define $N := \sum_{s\in G} f(X_s)F\circ\theta_s\epsilon_s$. It is easily checked that N is a HRM of X and using (11.5v) that its extension, N^*, to W defined in (8.18) is given by $N^* = \sum_{s\in G^*} f(Y_s)F\circ\theta_s\epsilon_s$. Using (8.21) we have for $\varphi \in p\mathcal{B}$

$$(11.8)\quad Q_m \sum_{s\in G^*} \varphi(s) f(Y_s) F\circ\theta_s = Q_m \int \varphi(s) N^*(ds)$$

$$= \int \varphi(t)\,dt\,\nu_N^m(1)\,.$$

But by (11.4)

$$U_N^1 1(x) = P^x \int_0^\infty e^{-s} N(ds)$$

$$= P^x \int_0^\infty e^{-s} f(X_s)\hat{P}^{X(s)}(F)\,dA_s = U_A^1(fg)(x)$$

where $g(x) = \hat{P}^x(F)$. Consequently (8.10) and (8.13) imply

$$\nu_N^m(1) = L^1(m, U_N^1 1) = L^1(m, U_A^1(fg)) = \nu(fg)$$
$$= \int f(x)\hat{P}^x(F)\nu(dx).$$

Combining this with (11.8) gives (11.7) for H of the form $\varphi \otimes f \otimes F$. Using (11.3i) and the σ-finiteness of ν, one sees that there exists an $H > 0$ of this form for which the two sides of (11.7) are equal and finite. The desired conclusion now follows by standard arguments. ∎

We are now prepared to formulate and prove Fitzsimmons' theorem.

(11.9) Theorem. *Let T be an exact terminal time of X and $m \in \text{Exc}$. Suppose that $P^m(0 < T \leq t) < \infty$ for some $t > 0$. Define a measure ν_τ^i on E by $\nu_\tau^i := Q_{m_i}[Y_\tau \in \cdot\,; 0 < \tau < 1]$ where τ is the extension of T to W and m_i is the invariant part of m. Let $\varphi \in p\mathcal{B}^+$ be bounded on $[0,t]$ for each $t < \infty$ and $f \in pb\mathcal{E}^*$. If $\int_0^\infty \varphi(t)\,dt = \infty$, then*

$$\text{(11.10)} \quad \lim_{t\to\infty} \left(\int_0^t \varphi(s)\,ds\right)^{-1} P^m[\varphi(T) f \circ X_T; 0 < T \leq t]$$
$$= \nu_\tau^i(f).$$

If $\varphi(t) \to L < \infty$ as $t \to \infty$, then for each $a > 0$

$$\text{(11.11)} \quad \lim_{t\to\infty} P^m[\varphi(T) f \circ X_T; t < T < t + a] = aL\nu_\tau^i(f).$$

Remark. Let $K := 1_{\mathbb{R}}(\tau)\epsilon_\tau$. Then K is a HRM of Y and $\nu_\tau^i = \nu_K^{m_i}$ is the Revuz measure of K relative to m_i defined in (8.25).

Proof. For the moment suppose only that $\varphi \in p\mathcal{B}^+$ and $f \in p\mathcal{E}^*$. Then

$$P^m[\varphi(T) f \circ X_T : \ 0 < T < \infty]$$
$$= Q_m[(\varphi(T) f \circ X_T 1_{\{0<T<\infty\}}) \circ \theta_0; \ \alpha < 0].$$

Observe that

$$(11.12) \quad (\varphi(T) f \circ X_T 1_{\{0<T<\infty\}}) \circ \theta_0 1_{\{\alpha<0\}}$$
$$= \sum_{s \in G^*, s \leq 0} \varphi(s + T \circ \theta_s) f \circ X_T \circ \theta_s 1_{\{0 < s + T \circ \theta_s < \infty\}}$$
$$+ \varphi(\tau) f \circ Y_\tau 1_{\{\alpha < 0 < \tau < \infty\}}.$$

Using (11.6) and defining $\hat{P}^\nu := \int \nu(dx) \hat{P}^x$ where $\nu = \nu_A^m$ as in (11.6), the Q_m expectation of the first term on the right hand side of (11.12) equals

$$\int_{-\infty}^0 dt\, \hat{P}^\nu[\varphi(t+T) f \circ X_T; \ 0 < t + T < \infty]$$
$$= \hat{P}^\nu \left[f \circ X_T \int_0^T \varphi(t)\, dt; \ T < \infty \right].$$

To compute the Q_m expectation of the second term on the right hand side of (11.12) we write $Q_m = Q_{m_i} + Q_{m_p}$ where $m = m_i + m_p$ is the decomposition of m into its invariant and purely excessive parts and compute each part separately. Since $\alpha = -\infty$ a.s. Q_{m_i} and $f \circ Y_\tau$ is invariant, using (9.8)—take both S and T in (9.8) equal to τ —we have

$$Q_{m_i}[\varphi(\tau) f \circ Y_\tau; \ 0 < \tau < \infty] = \nu_\tau^i(f) \int_0^\infty \varphi(t)\, dt.$$

Finally let $(\eta_t)_{t>0}$ be the entrance law such that $m_p = \int_0^\infty \eta_t \, dt$. Then from (6.11)

$$
\begin{aligned}
Q_{m_p}&[\varphi(\tau) f \circ Y_\tau;\ \alpha < 0 < \tau < \infty] \\
&= \int_{-\infty}^{\infty} Q_\eta[\varphi(\tau - t) f \circ Y_\tau;\ \alpha < t < \tau < \infty] \, dt \\
&= Q_\eta \left[f \circ Y_\tau \int_0^\tau \varphi(t) \, dt;\ \tau < \infty \right],
\end{aligned}
$$

where we have used $\alpha = 0$ a.s. Q_η to obtain the last equality. Combining these calculations gives the following formula:

$$
\begin{aligned}
(11.13)\quad P^m[\varphi(T) f \circ X_T;\ 0 < T < \infty] &= \nu^i_\tau(f) \int_0^\infty \varphi(t) \, dt \\
&+ \hat{P}^\nu \left[f \circ X_T \int_0^T \varphi(t) \, dt;\ T < \infty \right] \\
&+ Q_\eta \left[f \circ Y_\tau \int_0^\tau \varphi(t) \, dt;\ \tau < \infty \right],
\end{aligned}
$$

for $\varphi \in p\mathcal{B}^+$ and $f \in p\mathcal{E}^*$. Formula (11.13) does not require the condition $P^m[0 < T \le t] < \infty$ for some $t > 0$.

We now use (11.13) to prove (11.9). For (11.10), apply (11.13) with φ replaced by $\varphi 1_{]0,t]}$ to obtain

$$
\begin{aligned}
(11.14)\quad P^m[\varphi(T) f \circ X_T;\ 0 < T \le t] &= \int_0^t \varphi(s) \, ds \cdot \nu^i_\tau(f) \\
&+ \hat{P}^\nu \left[f \circ X_T \int_0^{T \wedge t} \varphi(s) \, ds;\ T < \infty \right] \\
&+ Q_\eta \left[f \circ Y_\tau \int_0^{\tau \wedge t} \varphi(s) \, ds;\ \tau < \infty \right].
\end{aligned}
$$

The hypothesis that $P^m[0 < T \le t] < \infty$ for one $t > 0$ implies that it is finite for all $t > 0$, and so the hypotheses on φ and f imply that the left hand side of (11.14) is finite for each t. But $\int_0^{T\wedge t} \varphi(s)\,ds \left(\int_0^t \varphi(s)\,ds\right)^{-1}$ and $\int_0^{\tau\wedge t} \varphi(s)\,ds \left(\int_0^t \varphi(s)\,ds\right)^{-1}$ decrease to zero as $t \to \infty$ on $\{T < \infty\}$ and $\{\tau < \infty\}$ respectively. Therefore dividing (11.14) by $\int_0^t \varphi(s)\,ds$ and letting $t \to \infty$ establishes (11.10). If $\varphi(t) \to L < \infty$ as $t \to \infty$, then φ is bounded and $\int_t^{a+t} \varphi(s)\,ds \to aL$ as $t \to \infty$. Write (11.13) with φ replaced by $1_{]t,t+a[}\,\varphi$. Using $P^m[0 < T \le t] < \infty$ for each t and the fact that $[(t+a) \wedge r - t] \vee 0$ decreases to zero as $t \to \infty$ for each $r \in]0,\infty[$, one easily obtains (11.11). ∎

Suppose that $B \in \mathcal{E}^e$. Since $\alpha \le \tau_B$, (6.7) implies that if B is m-cotransient, then it is m_i-strongly cotransient. Such a B is an m_i-equilibrium set and according to (10.32), $\gamma_B^i := \gamma_B^{m_i} = Q_{m_i}[Y_{\tau_B} \in \cdot\,;\ 0 < \tau_B < 1]$. Thus we have established the following corollary.

(11.15) Corollary. *Suppose B is m-cotransient and that $P^m[0 < T_B \le t] < \infty$ for some $t > 0$. Then with φ and f satisfying the hypotheses for* (11.10),

$$\text{(11.16)}\quad \lim_{t\to\infty} P^m[\varphi(T_B) f \circ X_{T_B};\ 0 < T_B \le t] \left(\int_0^t \varphi(s)\,ds\right)^{-1} = \gamma_B^i(f),$$

while under the hypotheses of (11.11),

$$\text{(11.17)}\quad \lim_{t\to\infty} P^m[\varphi(T_B) f \circ X_{T_B};\ t < T_B < t+a] = aL\gamma_B^i(f),$$

where $\gamma_B^i = \gamma_B^{m_i}$ is the capacitary measure of B relative to m_i.

Remark. The special case $\varphi = 1$ in (11.16) is often referred to as Spitzer's formula. See [**GSt86**] and [**St87**]. In [**St87**],

Steffens first proved (11.17) (when $\varphi = 1$) and then easily deduced (11.16) as the "Cesaro" mean of (11.17).

We can now identify the measure ν_B appearing in (10.44) in terms of an exit system $(\hat{P}^x, A)$ for T_B.

(11.18) Proposition. *Let $B \in \mathcal{E}^e$ be finely closed and let $(\hat{P}^x, A)$ be an exit system for T_B. Given $m \in \mathrm{Exc}$, let $\nu = \nu_A^m$ and $\Gamma^q = \Gamma_m^q$ for $q \geq 0$. Then*

$$\Gamma^q(B) = \Gamma(B) + qm(B) + \int_0^\infty (1 - e^{-qt})\nu_B(dt)$$

where $\nu_B(dt) = \hat{P}^\nu(T_B \in dt;\ T_B < \infty)$.

Proof. As in the proof of (10.44)

$$\Gamma^q(B) = \Gamma(B) + qm(B) + qP^{R_B m}(e^{-qT_B};\ T_B > 0).$$

But the last term above equals

$$\begin{aligned} &qQ_m[P^{Y(0)}(e^{-qT_B};\ T_B > 0);\ \tau_B < 0] \\ &\quad = qQ_m[e^{-qT_B \circ \theta_0};\ \tau_B < 0;\ 0 < T_B \circ \theta_0 < \infty] \\ &\quad = qQ_m\left[\sum_{s \in G^*, s \leq 0} e^{-q(s + T_B \circ \theta_s)} 1_{\{0 < s + T_B \circ \theta_s < \infty\}}\right], \end{aligned}$$

since $Q_m(\tau_B = 0) = 0$. Using (11.7) this becomes

$$\begin{aligned} &q\int_{-\infty}^0 dt\, \hat{P}^\nu[e^{-q(t+T_B)};\ 0 < t + T_B < \infty] \\ &\quad = \hat{P}^\nu[1 - e^{-qT_B};\ T_B < \infty] = \int_0^\infty (1 - e^{-qt})\nu_B(dt). \quad \blacksquare \end{aligned}$$

Perhaps we should remark that although neither $\hat{P}^x$ nor A is unique, the measure $\hat{P}^\nu$ on $(\Omega, \mathcal{F}^*)$ is uniquely determined by B and m, and hence so is ν_B.

Up to this point in the present section we have been assuming that X is a Borel right process. This is used only to insure that an exit system $(\hat{P}^x, A)$ for T satisfying (11.4) exists. However, the proof in [**S**, §74] of the existence of an exit system is carried out relative to a fixed P^x. Suppose X is a right process satisfying (6.2). Given an initial probability measure μ, let E_μ be as in (6.2ii). Then one may repeat the proof in [**S**, §74] to obtain a kernel $\hat{P}^x$ from $(E_\mu, \mathcal{E}_\mu^*)$ to $(\Omega, \mathcal{F}^*)$ and an additive functional A with a bounded 1-potential such that (11.4) holds with P^x replaced by P^μ. (Note that $\{\omega: X_t(\omega) \in E_\mu$ for all $t\} \in \mathcal{F}^*$ since E_μ is Lusinian. The construction in [**S**, §74] shows that A —it is called B in [**S**]—does not depend on μ, but a priori $\hat{P}^x$ may.) Now given $m \in \mathrm{Exc}$ by choosing μ a probability equivalent to m one obtains a system $(\hat{P}^x, A)$ possibly depending on m but such that (11.7) holds. (In the proof one only obtains $U_A^1 g \le U^1 h$ a.e. m and $U_N^1 1 = U_A^1(fg)$ a.e. m, but this suffices.) Hence Theorem 11.9, its proof, and corollaries are valid for any right process satisfying (6.2).

We close this section with two more applications of exit systems. We shall continue to suppose that X is a Borel right process, but in view of the preceding remarks these results are valid for any right process satisfying (6.2). The first application comes from [**FM86**] and expresses the operator R_T in terms of an exit system for T. For its statement we fix an exact terminal time T and an exit system $(\hat{P}^x, A)$ for T. Let $\mathrm{reg}(T) := \{x: P^x(T = 0) = 1\}$ be the set of regular points for T and define for $f \in p\mathcal{E}^*$

$$\hat{V} f(x) := \hat{P}^x \int_0^T f \circ X_t \, dt. \tag{11.19}$$

Then $\mathrm{reg}(T) \in \mathcal{E}^e$ and $\hat{V}$ is a kernel from $(E, \mathcal{E}^*)$ to $(E, \mathcal{E}^*)$.

(11.20) Proposition. *Let $m \in \mathrm{Exc}$. Then*

$$R_T m(f) = m(f\, 1_{\mathrm{reg}\,(T)}) + \nu \hat{V}(f)$$

where $\nu = \nu_A^m$ is the Revuz measure of A relative to m.

Proof. As usual τ denotes the extension of T to W. If $\alpha < 0 < \beta$, then $\{T \circ \theta_0 = 0\} \subset \{\tau \leq 0\}$. Since $Q_m(\tau = 0) = 0$, it follows that a.s. Q_m, $\{\tau < 0, T \circ \theta_0 = 0\} = \{T \circ \theta_0 = 0\}$. Therefore given $f \in p\mathcal{E}^*$,

$$Q_m[f \circ Y_0;\ \tau < 0;\ T \circ \theta_0 = 0]$$
$$= Q_m[f \circ Y_0 P^{Y(0)}(T = 0)] = m(f\, 1_{\mathrm{reg}\,(T)}).$$

On the other hand

$$Q_m[f \circ Y_0;\ \tau < 0,\ T \circ \theta_0 > 0]$$
$$= Q_m \sum_{s \in G^*, s \leq 0} f(X_{-s}) \circ \theta_s\, 1_{\{s + T \circ \theta_s > 0\}}$$
$$= \int_{-\infty}^{0} dt\, \hat{P}^\nu[f \circ X_{-t};\ 0 < t + T]$$
$$= \hat{P}^\nu \int_0^T f \circ X_t\, dt = \nu \hat{V}(f).$$

Since $T \circ \theta_0 \geq 0$, (11.20) follows from the above calculations and (7.9). ■

Our final application of exit systems is a generalization of a result of Kanda [**Ka78**]. It is another unpublished result of Fitzsimmons. For its statement we recall that set $B \in \mathcal{E}^e$ is m-semipolar provided there exists a semipolar set $\tilde{B}$, such that $B \Delta \tilde{B}$ is m-polar. Moreover B is m-semipolar if and only if

$$P^m(X_t \in B \quad \text{for uncountably many} \quad t) = 0,$$

and this in turn is evidently equivalent to

$$Q_m(Y_t \in B \quad \text{for uncountably many} \quad t) = 0.$$

See, for example, [**GS84**, §6]. We fix $m \in \mathrm{Exc}$.

(11.21) Theorem. *The following statements are equivalent.*

(i) *A set* $B \in \mathcal{E}^e$ *is* m*-semipolar if and only if it is* m*-polar.*

(ii) *If* $B \in \mathcal{E}^e$ *is not* m*-polar, then* $\sup_q \Gamma^q(B) = \infty$.

(iii) *If* K *is a compact totally thin set which is not* m*-polar, then* $\sup_q \Gamma^q(K) = \infty$.

Remarks. Of course, $\Gamma^q = \Gamma^q_m$ in (ii) and (iii). Recall that a set B is totally thin provided $\sup_{x \in B} P^x(e^{-T_B}) < 1$ and that every semipolar set is a countable union of totally thin sets. Moreover, if B is totally thin then $\{t: X_t \in B\}$ is discrete a.s. and consequently $\{t: Y_t \in B\}$ is discrete a.s. Q_m. If B is m-polar, $\Gamma^q(B) = 0$ for all $q \geq 0$ and so (ii) may be re-phrased as $\sup_q \Gamma^q(B)$ is either 0 or $+\infty$. A similar comment applies to (iii). Since $q \to \Gamma^q(B)$ is increasing the suprema in (ii) and (iii) may be replaced by limits as $q \to \infty$.

Proof. Since (i) holds for X if and only if it holds for X^q, the q-subprocess of X, it suffices to prove (11.21) for $m \in \text{Dis}$ by passing to X^q with $q > 0$ if necessary. Thus we suppose $m \in \text{Dis}$. Since (ii)$\Longrightarrow$(iii) is obvious we shall show that (i)$\Longrightarrow$(ii) and (iii)$\Longrightarrow$(i).

(i)$\Longrightarrow$(ii). Let $B \in \mathcal{E}^e$ with $\sup_q \Gamma^q(B) < \infty$. We may suppose that B is finely closed. It follows from (11.18) that $\Gamma(B) < \infty$, $m(B) = 0$, and $\nu_B(1) < \infty$. Hence, using (11.18) and (11.7), one has for each $k \in \mathbb{Z}$

$$\infty > \nu_B(1) = \hat{P}^\nu(T_B < \infty) = Q_m \sum_{s \in G^*,\, k < s \leq k+1} 1_{\{T_B \,\circ\, \theta_s < \infty\}} .$$

Consequently a.s. Q_m there are only a finite number of intervals contiguous to $D := \{t: Y_t \in B\}^-$ contained in any compact interval. Since $m(B) = 0$, the Fubini theorem implies that a.s. Q_m, $\{t: Y_t \in B\}$ has Lebesgue measure zero. But $\{t: Y_t \in B\}$ is a.s. Q_m right closed because B is finely closed, and so D

has Lebesgue measure zero a.s. Q_m. However a closed subset J of a compact interval I with $\ell(J) = 0$ and $I \setminus J$ consisting of a finite number of maximal disjoint open (in I) intervals must be finite. (J contains no non-void interval and if it is infinite it has a limit point t_0. Then there exist $t_n \downarrow\downarrow t_0$ (or $t_n \uparrow\uparrow t_0$) with $(t_n) \subset J$ which contradicts the hypothesis on $I \setminus J$.) Therefore D and hence $\{t : Y_t \in B\}$ is countable a.s. Q_m. Hence B is m-semipolar and then (i) implies that B is m-polar.

(iii)$\Longrightarrow$(i). Suppose (i) does not hold. Then there must exist a totally thin set B that is not m-polar. Since B is totally thin, B is finely closed and $m(B) = 0$. We shall reduce B several times in order to arrive at a set K which contradicts (iii). First of all because $m \in \mathrm{Dis}$ there exists $g > 0$ with $m(g) < \infty$ and $Ug \le 1$ a.e. m. See (2.17). Let V be the potential operator of X killed at time T_B so that $Vg(x) = P^x \int_0^{T_B} g \circ X_t \, dt$. Since B is totally thin, $T_B > 0$ a.s. and so $0 < Vg \le Ug$. Given $a > 0$ let $D := \{Vg \ge a\} \cap B$. Since D is totally thin and hence, finely closed, $X_{T_D} \in D$ a.s. on $\{T_D < \infty\}$. Let $T_0 = 0$, $T_1 = T_D, \cdots, T_{n+1} = T_n + T_D \circ \theta_{T_n}, \cdots$ be the *iterates* of T_D. Then

$$Ug(x) \ge P^x \sum_{n \ge 1} \int_{T_n}^{T_{n+1}} g \circ X_t \, dt$$

$$= \sum_{n \ge 1} P^x[Vg \circ X_{T_n};\ T_n < \infty] \ge a \sum_{n \ge 1} P^x(T_n < \infty).$$

Since $\{Vg \ge a\} \uparrow E$ as $a \downarrow 0$ we may choose a small enough that D is not m-polar. Thus D is a totally thin set, D is not m-polar, and D satisfies

$$\text{(11.22)} \quad P^x \left(\sum_{n \ge 1} 1_{\{T_n < \infty\}} \right) = \sum_{n \ge 1} P^x(T_n < \infty) \le c\, Ug(x)$$

where $c < \infty$ and $m(g) < \infty$. In particular, $P_D 1(x) \leq c\,Ug(x)$ and so $\Gamma(D) = L(m, P_D 1) \leq c\,m(g) < \infty$.

Next choose a probability μ equivalent to m and define a measure λ $(E, \mathcal{E}^e)$ by

$$\lambda(F) = \sum_{n \geq 1} P^{\mu}(X_{T_n} \in F;\ T_n < \infty),$$

where the T_n are the iterates of T_D. Since $Ug \leq 1$ a.e. m, (11.22) implies that $\lambda(E) < \infty$. Because λ is carried by D we may choose a compact subset K of D with $\lambda(K) > 0$. Thus K is not m-polar, K is totally thin, $\Gamma(K) < \infty$, and K satisfies (11.22); that is, $\sum_{n \geq 1} P^x(T_n < \infty) \leq c\,Ug(x)$ where the T_n are now the iterates of T_K. We shall show that $\sup_q \Gamma^q(K) < \infty$ contradicting (iii),

To this end first note that since K is totally thin one easily checks that an exit system for T_K is given by $(\hat{P}^x, A)$ where $\hat{P}^x := P^x$ for all x and $A_t := \sum_{n \leq 1} 1_{\{T_n \leq t\}}$, the T_n being the iterates of T_K. Since $\Gamma(K) < \infty$ and $m(K) = 0$, it is clear from (11.18) that $\sup_q \Gamma^q(K)$ is finite if and only if ν_K is a finite measure. But using the above exit system, (11.18) implies that $\nu_K(1) = \nu(P_K 1) \leq \nu(1)$ where ν is the Revuz measure of A. Now $U_A 1(x) = P^x \sum_{n \geq 1} 1_{\{T_n < \infty\}} \leq c\,Ug(x)$. Therefore

$$\nu(1) = L(m, U_A 1) \leq c\,m(g) < \infty.$$

Hence $\sup_q \Gamma^q(K) < \infty$. ∎

(11.23) Remarks. If m is a reference measure then an m-polar (resp. m semipolar) set is polar (resp. semipolar). Therefore if m is an excessive reference measure, $\sup_q \Gamma_m^q(K)$ is zero or infinite for all compact sets K is a necessary and sufficient condition that every semipolar set be polar. The condition that semipolar sets are polar is Hunt's celebrated hypothesis (H). The proof of Theorem 11.21 actually shows that if $\sup_q \Gamma^q(B) < \infty$, then B is m-semipolar, and that if B is m-semipolar but not m-polar, then B contains a non-m-polar compact subset K with $\sup_q \Gamma^q(K) < \infty$.

Appendix A

The purpose of this appendix is to give a proof of Meyer's master perfection theorem [**Me74**]. This is stated in [**S**, (60.5)], but not proved there. Our starting point is the perfection theorem (55.19) in [**S**]. Since there are several misprints in the proof of this result in [**S**] we shall make several comments on the proof before proceeding to Meyer's result. For the next few paragraphs the reader should have a copy of [**S**] at hand.

We begin by recalling the basic set-up. As in §1, we fix a right process X. For the moment we make no additional assumptions on the underlying sample space, Ω. We suppose given a decreasing weak raw multiplicative functional (WRMF), $m = (m_t(\omega))_{t \geq 0}$, of X as defined in [**S**, (54.2)]. For simplicity we take the σ-algebra $\mathcal{G}$ in [**S**, §55] to be $\mathcal{F}$. Thus $m_t \in \mathcal{F}$ and m satisfies [**S**, (54.1)]—the multiplicative property, $m_{t+s} = m_t m_s \circ \theta_t$ a.s. for each $t, s \geq 0$. In addition, we assume, as we may, that m satisfies the conditions in [**S**, (55.2)]:

(A.1) (i) $t \to m_t(\omega)$ is decreasing, right continuous, and takes values in $[0,1]$ for all $\omega \in \Omega$;

(ii) $m_0(\omega) = 0$ or 1 for all ω.

Throughout this appendix all MF's *are assumed to satisfy the conditions* (A.1). This hypothesis will not be repeated at each stage. It follows from (A.1i) that $(t, \omega) \to m_t(\omega)$ is $\mathcal{B}^+ \otimes \mathcal{F}$ measurable. We define

$$\text{(A.2)} \quad m(s,t,\omega) := m_{t-s}(\theta_s \omega); \quad \omega \in \Omega, \quad 0 \leq s \leq t < \infty.$$

Sharpe [**S**, (55.3)] defines $m_{]s,t]}(\omega) = m_{t-s}(\theta_s\omega)$, but we prefer $m(s,t,\omega)$ for typographical reasons. Of course, we often shall write $m(s,t)$ in place of $m(s,t,\cdot)$. The function $m(s,t)$ satisfies the condition [**S**, (55.4)] which we do not repeat. As in [**S**, (55.5)] we define, for $\omega \in \Omega$ and $0 \le s < t < \infty$,

$$\text{(A.3)} \qquad \psi_{s,t}(\omega) := \operatorname*{ess\,lim\,sup}_{r \downarrow s} m(r,t,\omega),$$

where for any numerical function $f(r)$ defined on $]s, s+\delta[$ for some $\delta > 0$

$$\operatorname*{ess\,lim\,sup}_{r\downarrow s} f(r) := \downarrow \lim_{\epsilon \downarrow 0} \operatorname*{ess\,sup}_{s<r<s+\epsilon} f(r).$$

See [**DM**, IV-36], for example. Of course, the essential supremum is taken relative to Lebesgue measure, ℓ, on $\mathbb{R}$. On pages 264–266 Sharpe develops the properties of ψ leading to the proof of Theorem 55.19. We shall not repeat the argument, but we shall make some comments about it. All page references are to [**S**].

The discussion on page 264 beginning with line 4 and leading to (55.6) is incorrect and should be ignored. The only thing that is clear at this point is that because of (A.1i), $t \to \psi_{s,t}(\omega)$ is decreasing on $]s,\infty[$ for each ω.

We repeat the definition of Ω_0 in (55.7) since it is critical for our later development. As in [**S**] we let ℓ denote Lebesgue measure on $\mathbb{R}^+$. Then $\Omega_0 \subset \Omega$ is the set of all $\omega \in \Omega$ satisfying:

(A.4) (i) the map $(s,t) \to m(s,t,\omega)$ defined on $\{(s,t) : 0 \le s \le t\} \subset \mathbb{R}^+ \times \mathbb{R}^+$ is $\ell \times \ell$ measurable;

(ii) $m(r,s,\omega)m(s,t,\omega) = m(r,t,\omega)$ for Lebesgue a.a. triples (r,s,t) with $0 \le r \le s \le t$.

Clearly $\omega \in \Omega_0 \Longrightarrow \theta_t\omega \in \Omega_0$ for each $t \ge 0$. On page 264, (55.9) states that $P^\mu(\Omega \backslash \Omega_0) = 0$ for all initial measures μ on

E. In particular, $\Omega_0 \in \mathcal{F}_0$ because $\mathcal{F}_0$ contains $\mathcal{N} := \{\Lambda \subset \Omega: P^\mu(\Lambda) = 0 \quad \text{for all} \quad \mu\}$.

In line -8 on page 265 replace the first occurrence of $m_{]r,t]}$ by $m_{]r,s]}$. The proof of (55.17) contains several misprints and so I suggest replacing lines 5 through 10 on page 266 (beginning with "However" to the end of the proof of (55.17)) with the following:

Let $s' = \sup\{s < t: \psi_{r,s} = 0 \ \forall r < s\}$ and set $s' = t$ if there are no such s. If $s' < s < t$, then there exists an $r < s$ with $\psi_{r,s} > 0$ and so $m(s,t) = \psi_{s,t}$ for a.a. $s \in]s', t[$. If $s < s'$, $\psi_{r,s} = 0$ for all $r < s$ and hence by (55.16), $\psi_{r,t} = 0$ for all $r < s$. Therefore $\psi_{s,t} = 0$ for all $s < s'$. If $s < s'$ and for some $\epsilon > 0$, $A(\epsilon) := \{r \in]0, s[: m(r,s) \geq \epsilon\}$ has positive Lebesgue measure, then there exists a point $r_0 \in A(\epsilon)$ at which the density of $A(\epsilon)$ from the right is one. But this implies that $\psi_{r_0,s} \geq \epsilon$ contradicting $s < s'$. Consequently $m(r,s) = 0$ for a.a. $r \in]0, s[$. But $s \to m(r,s)$ is decreasing, so $m(r,t) = 0$ for a.a. $r \in]0, s[$, and hence for a.a. $r \in]0, s'[$. Combining these remarks $\psi_{s,t} = m(s,t)$ for a.a. $s < t$ as claimed.

It follows from (55.16) that $s \to \psi_{s,t}(\omega)$ is increasing on $[0, t[$ for each $\omega \in \Omega_0$. Fix $\omega \in \Omega_0$ and suppress it. Then $\psi_{s,t} \leq \psi_{s+,t}$ for $0 \leq s < t$. Fix $0 \leq s < t$ and suppose $\psi_{s,t} < \delta$. Then there exists $\epsilon > 0$ with $s + \epsilon < t$ such that $m(r,t) < \delta$ for a.a. $r \in]s, s+\epsilon[$. Hence $\psi_{u,t} \leq \delta$ for $u \in]s, s+\epsilon[$, and so $\psi_{s+,t} \leq \delta$. It now follows that $s \to \psi_{s,t}$ is increasing and right continuous on $[0, t[$. This is the assertion (55.6) but only for $\omega \in \Omega_0$.

Now define $R(\omega) := \inf\{t: \theta_t\omega \in \Omega_0\}$ as in Sharpe just above (55.18). It follows from (55.8) that $\{R = 0\} = \Omega_0$ and that $\theta_t\omega \in \Omega_0$ for all $t \in [R(\omega), \infty[$. (Recall $\theta_0\omega = \omega$ for all $\omega \in \Omega$.) In addition, $R \in \mathcal{F}_0$ since $R = 0$ a.s. For $\omega \in \Omega_0$, define $\psi_{0,0}(\omega) = \uparrow \lim_{t \downarrow 0} \psi_{0,t}(\omega)$. Finally for $t \geq 0$, define

(A.5)

$$n_t(\omega) = \begin{cases} \psi_{0,t}(\omega) & \text{if} \quad \omega \in \Omega_0 \\ 1 & \text{if} \quad t < R(\omega) \\ \psi_{0,t-R(\omega)}(\theta_{R(\omega)}\omega) & \text{if} \quad \omega \notin \Omega_0 \quad \text{and} \quad t \geq R(\omega) \end{cases}$$

This is (55.18) but with the definition for $t = 0$ or $t = R(\omega)$ made explicit and we have used n rather than $\bar{m}$ as in [**S**] for notational simplicity. Of course, $t < R(\omega)$ implies that $\omega \notin \Omega_0$. The following theorem is a re-statement of [**S**, (55.19)].

(A.6) **Theorem.** *Let m be a* WRMF *as above. Define $n = (n_t)_{t\geq 0}$ by* (A.5). *Then n is a perfect exact* RMF *of X such that:*

- (i) *$\forall\omega \in \Omega$, $s \to n_{t-s}(\theta_s\omega)$ is right continuous and increasing on $[0,t[$; $t > 0$;*
- (ii) *$\forall\omega \in \Omega$, $t \to n_{t-s}(\theta_s\omega)$ is right continuous and decreasing on $[s,\infty[$, $s \geq 0$;*
- (iii) *$m_t = n_t$ for all $t \geq 0$* a.s. *on $\{m_0 = 1\}$.*
- (iv) *$m_t \leq n_t$ for all $t \geq 0$* a.s.*;*
- (v) *m and n are indistinguishable if m is exact;*
- (vi) *if m is $(\mathcal{F}_t)$ adapted, so is n.*

Condition (vi) does not appear in (55.19), but is an immediate consequence of (A.5) and the facts that R and Ω_0 are $\mathcal{F}_0$ measurable and that $\psi_{s,t} \in \mathcal{F}_t$ for $s < t$ when m is adapted. This last assertion follows from the definition (A.3) and [**DM**, IV-38]. We remind the reader that a WRMF, m, is *exact* provided for each $t > 0$ and sequence $t_n \downarrow\downarrow 0$ one has $m_{t-t_n} \circ \theta_{t_n} \to 0$ a.s. as $n \to \infty$. [**S**, (54.9)]. The proof of (55.19) in [**S**] does not mention (v). But the argument for (v) is the same as that for (iii). Of course, (A.6i) represents a considerable strengthening of the exactness of n.

This completes our comments on §55 in [**S**] and we now turn to the problem of improving the measurability of n. We fix an exact weak MF, m, of X. That is, m is an exact WRMF that is $(\mathcal{F}_t)$ adapted. See [**S**, (54.5)]. Then according to [**S**, (60.2)], m is indistinguishable from an exact WMF, m^e such that $m^e_t \in \mathcal{F}^e_{t+}$ for each $t \geq 0$ where $\mathcal{F}^e_t$ is the σ-algebra generated by $f \circ X_s$ for $s \leq t$ and $f \in \mathcal{E}^e$. (Sharpe only states this for MFs rather than WMFs. But the proof only uses the weak multiplicative functional property. Note that the reference (57.9) in the proof of (60.2) should be (56.9).) Consequently we may and do assume that m itself is an exact WMF which is $(\mathcal{F}^e_{t+})$ adapted. *We fix such an m for the remainder of this appendix.* For notational simplicity we set $\Gamma_t(\omega) = \psi_{0,t}(\omega)$ where $\psi_{0,t}$ is defined in (A.3); that is for $t > 0$

$$\text{(A.7)} \qquad \Gamma_t(\omega) := \operatorname*{ess\,lim\,sup}_{r \downarrow 0} m_{t-r}(\theta_r \omega).$$

(A.8) **Lemma.** *For each $t > 0$, $\Gamma_t \in \mathcal{F}^*$ and $\Omega_0 \in \mathcal{F}^*$.*

Proof. Fix $t > 0$ and a probability measure, P, on $(\Omega, \mathcal{F}^0)$. Define $\overline{P}$ on $(\Omega, \mathcal{F}^0)$ by $\overline{P} := \int_0^\infty e^{-u}\theta_u(P)\,du$. If $r \in \mathbb{Q}^+$, $r < t$, then $m_{t-r} \in \mathcal{F}^e_{(t-r)+} \subset \mathcal{F}^*$, and so there exist $\overline{H}_r$, $\overline{H}'_r \in \mathcal{F}^0$ with $0 \le \overline{H}_r \le m_{t-r} \le \overline{H}'_r \le 1$ and $\overline{P}(\overline{H}_r) = \overline{P}(\overline{H}'_r)$. If $u \in \mathbb{R}^+$, $0 < u \le t$, define

$$H_u = \liminf_{r \in \mathbb{Q}, r \uparrow u} \overline{H}_r, \qquad H'_u = \liminf_{r \in \mathbb{Q}, r \uparrow u} \overline{H}'_r.$$

Then H and H' are $\mathcal{B}_t \otimes \mathcal{F}^0$ measurable as functions on $]0,t] \times \Omega$ where $\mathcal{B}_t := \mathcal{B}(]0,t])$. See [**DM**, IV-17.2]. But $s \to m_{t-s}(\omega)$ is left continuous on $]0,t]$ for every ω by (A.1i) and so $H_u \le m_{t-u} \le H'_u$ for $0 < u \le t$. Clearly $\overline{P}(H_u) = \overline{P}(H'_u)$. Now $H_u \circ \theta_s \le m_{t-u} \circ \theta_s \le H'_u \circ \theta_s$, $0 < u \le t$ and $s \ge 0$. Since $(u,\omega) \to H_u(\omega)$ is $\mathcal{B}_t \otimes \mathcal{F}^0$ measurable, it follows that $(u,s,\omega) \to H_u(\theta_s\omega)$ is $\mathcal{B}_t \otimes \mathcal{B}^+ \otimes \mathcal{F}^0$ measurable where $\mathcal{B}^+ := \mathcal{B}(\mathbb{R}^+)$. Hence for each $0 < u \le t$, $P(H_u \circ \theta_s) = P(H'_u \circ \theta_s)$ for a.e. (Lebesgue) $s \ge 0$, because $\overline{P}(H_u) = \overline{P}(H'_u)$. Consequently by the Fubini theorem, for a.e. $s \ge 0$ and a.e. ω (P), $H_u(\theta_s\omega) = H'_u(\theta_s\omega)$ for a.e. $u \in]0,t]$. Define for $0 < s \le t$

$$\gamma_s(\omega) = \operatorname*{ess\,lim\,inf}_{u \uparrow s} H_u(\theta_s\omega); \; \gamma'_s(\omega) = \operatorname*{ess\,lim\,inf}_{u \uparrow s} H'_u(\theta_s\omega).$$

According to [**DM**, IV-38], γ and γ' are $\mathcal{B}_t \otimes \mathcal{F}^0$ measurable on $]0,t] \times \Omega$ and from the above, for P a.e. ω, $\gamma_s(\omega) = \gamma'_s(\omega)$ for a.e. $s \leq t$. Again (A.1i) tells us that $\gamma_s \leq m_{t-s} \circ \theta_s \leq \gamma'_s$ for $0 < s \leq t$ identically in ω. Finally defining $\gamma = \operatorname{ess}\liminf_{s \downarrow 0} \gamma_s$ and $\gamma' = \operatorname{ess}\liminf_{s \downarrow 0} \gamma'_s$, it follows that $\gamma \leq \Gamma_t \leq \gamma'$, $\gamma = \gamma'$ a.e. P, and that $\gamma, \gamma' \in \mathcal{F}^0$. Since P was an arbitrary probability on $(\Omega, \mathcal{F}^0)$, $\Gamma_t \in \mathcal{F}^*$.

We now turn to showing $\Omega_0 \in \mathcal{F}^*$. Let Ω_1 (resp. Ω_2) be the set of $\omega \in \Omega$ satisfying (A.4i) (resp. (A.4ii)). It suffices to show $\Omega_i \in \mathcal{F}^*$, $i = 1, 2$. For fixed t, $m_t \in \mathcal{F}^e_{t+} \subset \mathcal{F}^*$ and so $(u, \omega) \to m_t(\theta_u \omega)$ is in the completion of $\mathcal{B}^+ \otimes \mathcal{F}^0$ relative to $\ell \times P$ where P is any fixed probability on $(\Omega, \mathcal{F}^0)$. Since $t \to m_t(\theta_u \omega)$ is right continuous, it follows that $(t, u, \omega) \to m_t(\theta_u \omega)$ is in the $\ell \times \ell \times P$ completion of $\mathcal{B}^+ \otimes \mathcal{B}^+ \otimes \mathcal{F}^0$. Since $(s,t) \to (t-s, s)$ is linear we see that $(s, t, \omega) \to m_{t-s}(\theta_s \omega) = m(s, t, \omega)$ is also in the $\ell \times \ell \times P$ completion of $\mathcal{B}^+ \otimes \mathcal{B}^+ \otimes \mathcal{F}^0$. In particular, using the Fubini theorem, a.s. P, $(s,t) \to m_{t-s} \circ \theta_s$ is $\ell \times \ell$ measurable on $\{(s,t): 0 \leq s \leq t\}$. Hence Ω_1 is in the P-completion of $\mathcal{F}^0$ and $P(\Omega_1) = 1$. Since P is an arbitrary probability, $\Omega_1 \in \mathcal{F}^*$. (If $\mathcal{F}^0$ contains all singletons $\{\omega\}$, taking $P = \epsilon_\omega$ we see that $\Omega_1 = \Omega$ in this case.) For Ω_2, define

$$A_u(\omega) = \int_{C_u} 1_{\{m(r,s,\omega)m(s,t,\omega) \neq m(r,t,\omega)\}} \, dr \, ds \, dt$$

where $C_u = \{0 \leq r \leq s \leq t : r^2 + s^2 + t^2 \leq u^2\}$. Because of the measurability of $m(s,t,\omega)$ just established, $A_u \in \mathcal{F}^*$ and $u \to A_u(\omega)$ is continuous and finite on $[0, \infty[$. Now $T := \inf\{u : A_u > 0\} \in \mathcal{F}^*$ by [**DM**, III-44]—note $(\mathcal{F}^*)^* = \mathcal{F}^*$. But $\Omega_2 = \{T = \infty\}$ and so $\Omega_2 \in \mathcal{F}^*$. ∎

We now give a first preliminary improvement in the measurability of n.

(A.9) **Proposition**. *Let m be as above. Then n defined in (A.5) is such that $n_t \in \mathcal{F}^*$ for each $t \geq 0$.*

Proof. First note that if $F \in p\mathcal{F}^*$, then $(t, \omega) \to F(\theta_t\omega)$ is $(\mathcal{B}^+ \otimes \mathcal{F}^0)^*$ measurable. We may now write

$$R = \inf \left\{ t\colon \int_0^t 1_{\Omega_0}(\theta_s\omega)\, ds > 0 \right\},$$

from which it follows that R is $\mathcal{F}^*$ measurable. Therefore, in view of (A.7) and (A.8) it suffices to show for $t > 0$, that

$$G(\omega) := 1_{\{0 < R(\omega) \le t\}} \Gamma_{t-R(\omega)}(\theta_{R(\omega)}\omega)$$

is $\mathcal{F}^*$ measurable. But $\omega' = \theta_{R(\omega)}\,\omega \in \Omega_0$ if $R(\omega) < \infty$ and so $s \to 1_{\{R(\omega)<\infty\}}\Gamma_{t-s}(\theta_{R(\omega)}\omega)$ is left continuous on $]0, t[$. Now $\Gamma_{t-s} \in \mathcal{F}^*$ by (A.8) and so $(u, \omega) \to \Gamma_{t-s}(\theta_u\omega)$ is $(\mathcal{B}^+ \otimes \mathcal{F}^0)^*$ mesurable. Since $R \in \mathcal{F}^*$ it follows that $\omega \to \Gamma_{t-s}(\theta_{R(\omega)}\omega)$ is $\mathcal{F}^*$ measurable. Combining these observations yields $G \in \mathcal{F}^*$. ∎

At this point we shall assume that Ω is the canonical space of all right continuous paths from $[0, \infty[$ to $E \cup \{\Delta\}$ with Δ as cemetery point and $[\Delta](t) := \Delta$ for all $t \ge 0$. In this case each singleton $\{\omega\} \in \mathcal{F}^0$ and, as remarked in the preceding proof, this implies that (A.4i) is valid for every $\omega \in \Omega$. Let $k_t\colon \Omega \to \Omega$ be the killing operators defined for $t \ge 0$ by

$$k_t\omega(s) = \omega(s) \quad \text{if} \quad s < t; \; k_t\omega(s) = \Delta \quad \text{if} \quad s \ge t.$$

The assumption that Ω is the canonical space is just a convenience; the only things that we shall use are the fact that (A.4i) holds for all $\omega \in \Omega$ and the existence of maps $k_t\colon \Omega \to \Omega$, $t \ge 0$ with the following properties for every $\omega \in \Omega$. See also [**S**, (11.3)].

(A.10) (i) $X_s(k_t\omega) = X_s(\omega)$ if $s < t$, $X_s(k_t\omega) = \Delta$ if $s \ge t$;

(ii) $k_t k_s = k_{t\wedge s}$; $k_t([\Delta]) = [\Delta]$;

(iii) $k_t\theta_s = \theta_s k_{t+s}$;

(iv) $\theta_s\omega = [\Delta]$ if $s \geq \zeta(\omega)$, $k_0\omega = [\Delta]$, and $k_t\omega = \omega$ if $t > \zeta(\omega)$.

Note that (A.10i) implies that $\zeta \circ k_t = \zeta \wedge t$, and so if $s \geq t$, then $\theta_s k_t \omega = [\Delta]$ because of (A.10iv). It also follows from (A.10i) that $k_t \in \mathcal{F}_t^0 | \mathcal{F}^0$ and $k_t \in \mathcal{F}_t^e | \mathcal{F}^e$. Therefore $k_t \in \mathcal{F}_t^* | \mathcal{F}^*$. Moreover if $F \in \mathcal{F}^0$ (resp. $\mathcal{F}^e$), then $F \in \mathcal{F}_{t+}^0$ (resp. $\mathcal{F}_{t+}^e$) if and only if $F \circ k_s = F$ for each $s > t$. Hence (A.7) and the fact that $m_{t-r} \circ \theta_r \in \mathcal{F}_{t+}^e$ if $r < t$ imply that $\Gamma_t \circ k_s = \Gamma_t$ if $s > t$. But $\Gamma_t \in \mathcal{F}^*$ and so $\Gamma_t \in \mathcal{F}_{t+}^*$. We now summarize the properties of $(\Gamma_t;\, t > 0)$ that have been established so far for ease of reference:

(A.11) (i) $t \to \Gamma_t(\omega)$ is decreasing on $]0, \infty[$ with values in $[0, 1]$ for every ω.

(ii) $\Gamma_t \in \mathcal{F}_{t+}^*$, $t > 0$.

(iii) $\Gamma_t(k_s\omega) = \Gamma_t(\omega)$, $0 < t < s$, $\omega \in \Omega$.

(iv) $\Omega_0 \in \mathcal{F}^*$ where Ω_0 is defined in (A.4), and

(a) $\omega \in \Omega_0 \Longrightarrow \theta_t\omega \in \Omega_0$, $t \geq 0$;

(b) $P^\mu(\Omega_0) = 1$ for all probabilities μ on E;

(c) $s_n \downarrow\downarrow 0$, $\theta_{s_n}\omega \in \Omega_0$ for each $n \Longrightarrow \omega \in \Omega_0$;

(d) $t \to \Gamma_t(\omega)$ is right continuous on $]0, \infty[$ for $\omega \in \Omega_0$;

(e) $\Gamma_{t+s}(\omega) = \Gamma_t(\omega)\Gamma_s(\theta_t\omega)$, for $s, t > 0$ and $\omega \in \Omega_0$;

(f) $s \to \Gamma_{t-s}(\theta_s\omega)$ is increasing and right continuous on $[0, t[$ for $t > 0$ and $\omega \in \Omega_0$.

We now make two additional assumptions on m; the first of which is harmless and the second will eventually be weakened.

(A.12) $$m_t([\Delta]) = 1 \quad \text{for all} \quad t \geq 0.$$

(A.13) $$m_t = m_{\zeta-} \quad \text{if} \quad t \geq \zeta \quad \text{where} \quad m_{0-} := 1 .$$

Conditions (A.12) and (A.13) are in force until further notice.

(A.14) **Lemma.** *For each* $u \geq 0$, $\omega \in \Omega_0 \Longrightarrow k_u\omega \in \Omega_0$.

Proof. Since (A.4i) holds for all $\omega \in \Omega$ we must show that if $\omega \in \Omega_0$, then (A.4ii) holds for $k_u\omega$. If $u = 0$, this is immediate from (A.10iv) and (A.12). Using Fubini's theorem and the right continuity of $t \to m_t(\omega)$ one sees that under the present assumptions $\omega \in \Omega_0$ if and only if for each $t > 0$

(A.15) $$m_{t-r}(\theta_r\omega) = m_{s-r}(\theta_r\omega)m_{t-s}(\theta_s\omega)$$
$$\text{for a.a.} \quad r, s \quad \text{with} \quad 0 \leq r \leq s \leq t .$$

We now fix $\omega \in \Omega_0$ and $u > 0$. To show $k_u\omega \in \Omega_0$ we must show that (A.15) with ω replaced by $k_u\omega$ holds for each $t > 0$. Since $m_{t-s} \circ \theta_s$, $m_{s-r} \circ \theta_r$, and $m_{t-r} \circ \theta_r$ are in $\mathcal{F}^e_{t+}$ if $0 \leq r \leq s \leq t$, this is clear if $t < u$. Suppose $0 < u \leq t$. Let $z = \zeta(k_u\omega) \leq u$. If $z \leq r \leq s \leq t$, then $\theta_r k_u\omega = [\Delta]$ and $\theta_s k_u\omega = [\Delta]$, so (A.15) holds for $k_u\omega$ identically in this range. If $r < z \leq s$, then $t - r \geq s - r \geq z - r = \zeta(\theta_r k_u\omega)$, while $\theta_s k_u\omega = [\Delta]$. Thus (A.15) for $k_u\omega$ holds identically in this range because of (A.12) and (A.13). Finally we consider the range, $0 \leq r \leq s < z$. Fix v with $s < v < z$. Since $v < z \leq u$, $m_{v-r}(\theta_r k_u\omega) = m_{v-r}(\theta_r\omega)$, etc., and so

$$m_{v-r}(\theta_r k_u\omega) = m_{s-r}(\theta_r k_u\omega)m_{v-s}(\theta_s k_u\omega)$$

for a.a. $0 \leq r \leq s < v$. Letting $v \uparrow z$ gives

(A.16) $$m_{(z-r)-}(\theta_r k_u\omega) = m_{s-r}(\theta_r k_u\omega)m_{(z-s)-}(\theta_s k_u\omega)$$

for a.a. $0 \leq r \leq s < z$. But $z - r = \zeta(\theta_r k_u\omega)$ and $z - s = \zeta(\theta_s k_u\omega)$ if $0 \leq r \leq s < z$, and since $t - r \geq z - r$ and $t - s \geq z - s$, we conclude from (A.13) that (A.16) holds with $(z-r)-$ and $(z-s)-$ replaced by $t-r$ and $t-s$ respectively for a.a. $0 \le r \le s < z$. Combining the various cases we see that $k_u\omega \in \Omega_0$. ∎

We now are going to remove the exceptional set Ω_0 while preserving the other properties of $\Gamma = (\Gamma_t)_{t>0}$ listed in (A.11). To this end given $t > 0$ and $\omega \in \Omega$ call $\tau := \{(s_i, t_i)\}_{1 \le i \le n}$ a *good partition for* ω *of* $[0,t]$ (g.p. $\omega[0,t]$) provided:

(A.17) (i) $0 \le s_1 < t_1 \le s_2 < t_2 \le \cdots \le s_n < t_n \le t$;

(ii) $k_{u_i - s_i}\theta_{s_i}\omega \in \Omega_0$ for some $u_i > t_i$, $1 \le i \le n$.

If, in addition, each s_i and t_i is rational except for t_n in case $t_n = t$ we shall say that τ is *rational*; that is, τ is rational provided $\tau \cap [0, t[\subset \mathbb{Q}$. We shall use τ both for the collection of ordered pairs (s_i, t_i) and the set $\{s_i, t_i;\ 1 \le i \le n\}$. Given $t > 0$ and $\omega \in \Omega$, define

$$\text{(A.18)} \qquad J_t^\tau(\omega) := \prod_{i=1}^{n} \Gamma_{t_i - s_i}(\theta_{s_i}\omega), \qquad \tau \text{ a g.p. } \omega[0,t],$$

$$\text{(A.19)} \qquad J_t(\omega) = \inf_\tau J_t^\tau(\omega),$$

where the infimum in (A.19) is over all g.p. $\omega[0,t]$. Clearly $t \to J_t(\omega)$ is decreasing. Obviously $0 \le J_t(\omega) \le J_t^\tau(\omega) \le 1$. By convention the empty partition is a g.p. $\omega[0,t]$ for each $t > 0$ and ω, and $J_t^\tau(\omega) := 1$ if τ is empty.

(A.20) **Lemma.** *If* $\omega \in \Omega_0$ *and* $t > 0$, $\Gamma_t(\omega) = J_t(\omega)$.

Proof. By (A.14), $\tau = \{(0,t)\}$ is a g.p. $\omega[0,t]$ and so $J_t(\omega) \le \Gamma_t(\omega)$. On the other hand, given a g.p. $\omega[0,t]$, $\tau = \{(s_i, t_i)\}_{1 \le i \le n}$, using (A.11) we have

$$\begin{aligned}\Gamma_t(\omega) &= \Gamma_{s_1}(\omega)\Gamma_{t-s_1}(\theta_{s_1}\omega) \le \Gamma_{t-s_1}(\theta_{s_1}\omega)\\ &= \Gamma_{(t_1-s_1)+(t-t_1)}(\theta_{s_1}\omega) = \Gamma_{t_1-s_1}(\theta_{s_1}\omega)\Gamma_{t-t_1}(\theta_{t_1}\omega),\end{aligned}$$

and proceeding inductively, $\Gamma_t(\omega) \le J_t^\tau(\omega)$. Therefore $\Gamma_t(\omega) \le J_t(\omega)$. ∎

(A.21) **Lemma.** *$J_t(k_r\omega) = J_t(\omega)$ for all $\omega \in \Omega$ and all $0 < t < r$.*

Proof. First note that if $0 < s < u \le r$, then

$$k_{u-s}\theta_s k_r\omega = k_{u-s}k_{r-s}\theta_s\omega = k_{u-s}\theta_s\omega\,,$$

while if $0 < s < r < u$,

$$k_{u-s}\theta_s k_r\omega = k_{r-s}\theta_s\omega = k_{r-s}k_{u-s}\theta_s\omega\,.$$

Therefore if $0 < s < r \wedge u$, then, using (A.14), $k_{u-s}\theta_s\omega \in \Omega_0$ implies that $k_{u-s}\theta_s k_r\omega \in \Omega_0$, while $k_{u-s}\theta_s k_r\omega \in \Omega_0$ implies that $k_{(u\wedge r)-s}\theta_s\omega \in \Omega_0$. Consequently if $0 < t < r$, τ is a g.p. $\omega[0,t]$ if and only if τ is a g.p. $k_r\omega[0,t]$. Lemma (A.21) is now an immediate consequence of (A.11iii). ∎

(A.22) **Proposition.** *$J_{t+s}(\omega) = J_t(\omega)J_s(\theta_t\omega)$ for all $t, s > 0$ and $\omega \in \Omega$.*

Proof. Fix $s, t > 0$ and $\omega \in \Omega$. Let τ be a g.p. $\omega[0,t]$ and τ' a g.p. $\theta_t\omega[0,s]$. Then $\tau'' := \tau \cup (\tau' + t)$ is a g.p. $\omega[0, t+s]$. Given $0 < \epsilon < 1$ we may choose τ and τ' so that $J_t^\tau(\omega) \le J_t(\omega) + \epsilon$ and $J_s^{\tau'}(\theta_t\omega) \le J_s(\theta_t\omega) + \epsilon$. Therefore

$$J_{t+s}(\omega) \le J_{t+s}^{\tau''}(\omega) = J_t^\tau(\omega)J_s^{\tau'}(\theta_t\omega) \le J_t(\omega)J_s(\theta_t\omega) + 3\epsilon\,,$$

and so $J_{t+s}(\omega) \le J_t(\omega)J_s(\theta_t\omega)$. For the opposite inequality fix $\epsilon > 0$ and choose a g.p. $\omega[0, t+s]$, $\tau = \{(s_i, t_i);\ 1 \le i \le n\}$, so that $J_{t+s}(\omega) \le J_{t+s}^\tau(\omega) + \epsilon$. We now distinguish two cases.

Case 1. It is *not* the case that $s_k < t < t_k$ for some k, $1 \le k \le n$. If $t > t_n$, let $\tau' = \tau$ and τ'' be empty, while if $t < s_1$, let τ' be empty and $\tau'' = \tau - t$. If $t = s_k$ for some k, let $\tau' = \{(s_i, t_i);\ 1 \le i < k\}$ and $\tau'' = \{(s_i - t, t_i - t);\ k \le i \le n\}$, while if $t = t_k$ for some k let $\tau' = \{(s_i, t_i);$

$1 \le i \le k\}$ and $\tau'' = \{(s_i - t, t_i - t);\ k < i \le n\}$. Note that τ' is empty if $t \le s_1$ and τ'' is empty if $t \ge t_n$. Clearly τ' (resp. τ'') is a g.p. $\omega[0,t]$ (resp. a g.p. $\theta_t\omega[0,s]$). Therefore

$$J_t(\omega)J_s(\theta_t\omega) \le J_t^{\tau'}(\omega)J_s^{\tau''}(\theta_t\omega) = J_{t+s}^{\tau}(\omega) \le J_{t+s}(\omega) + \epsilon$$

and so $J_t(\omega)J_s(\theta_t\omega) \le J_{t+s}(\omega)$.

Case 2. $s_k < t < t_k$ for some k. In this case let

$$\tau' = \{(s_i, t_i);\ 1 \le i \le k-1\} \cup \{(s_k, t)\},$$
$$\tau'' = \{(0, t_k - t)\} \cup \{(s_i - t, t_i - t);\ k+1 \le i \le n\}.$$

By hypothesis there exists $u > t_k$ such that $k_{u-s_k}\theta_{s_k}\omega \in \Omega_0$. Since $t_k > t$ this means that τ' is a g.p. $\omega[0,t]$. Also

$$k_{u-t}\theta_t\omega = k_{u-t}\theta_{t-s_k}\theta_{s_k}\omega = \theta_{t-s_k}k_{u-s_k}\theta_{s_k}\omega \in \Omega_0$$

by (A.11iv-a). Since $u - t > t_k - t$, τ'' is a g.p. $\theta_t\omega[0,s]$. Now $u > t_k > t > s_k$ and so, using (A.11iii) and (A.11iv-e)

$$\begin{aligned}
\Gamma_{t_k - s_k}(\theta_{s_k}\omega) &= \Gamma_{t-s_k+(t_k-t)}(k_{u-s_k}\theta_{s_k}\omega) \\
&= \Gamma_{t-s_k}(k_{u-s_k}\theta_{s_k}\omega)\Gamma_{t_k-t}(\theta_{t-s_k}k_{u-s_k}\theta_{s_k}\omega) \\
&= \Gamma_{t-s_k}(\theta_{s_k}\omega)\Gamma_{t_k-t}(k_{u-t}\theta_t\omega) \\
&= \Gamma_{t-s_k}(\theta_{s_k}\omega)\Gamma_{t_k-t}(\theta_t\omega).
\end{aligned}$$

Consequently $J_{t+s}^{\tau}(\omega) = J_t^{\tau'}(\omega)J_s^{\tau''}(\theta_t\omega)$. Now arguing exactly as in Case 1 we obtain $J_t(\omega)J_s(\theta_t\omega) \le J_{t+s}(\omega)$. ∎

(A.23) **Lemma.** *Fix $\omega \in \Omega$ and $0 \le s < t$. If there exists a $v > t$ with $k_{v-s}\theta_s\omega \in \Omega_0$, then $u \to \Gamma_{t-u}(\theta_u\omega)$ is increasing and right continuous on $[s,t[$.*

Proof. If $u \in [s,t[$, then

$$\begin{aligned}
\Gamma_{t-u}(\theta_u\omega) &= \Gamma_{t-u}(k_{v-u}\theta_u\omega) = \Gamma_{t-u}(k_{v-u}\theta_{u-s}\theta_s\omega) \\
&= \Gamma_{t-u}(\theta_{u-s}k_{v-s}\theta_s\omega) = \Gamma_{t-s-(u-s)}(\theta_{u-s}\omega'),
\end{aligned}$$

where $\omega' = k_{v-s}\theta_s\omega \in \Omega_0$. The conclusion now follows from (A.11iv-f). ∎

(A.24) **Proposition.** *For each* $t > 0$ *and* $\omega \in \Omega$, $J_t(\omega) = \inf_\tau J_t^\tau(\omega)$ *where the infimum is over all rational g.p.* $\omega[0,t]$.

Proof. Let $r(t)$ be the collection of all rational g.p. $\omega[0,t]$. Trivially $J_t(\omega) \leq \inf\{J_t^\tau(\omega)\colon \tau \in r(t)\}$. For the opposite inequality fix $\epsilon > 0$ and choose a g.p. $\omega[0,t]$, $\tau = \{(s_i, t_i);\ 1 \leq i \leq n\}$, such that $J_t^\tau(\omega) \leq J_t(\omega) + \epsilon$. Using Lemma A.23, for each i we may choose $r_i \in]s_i, t_i[$ such that $r_i \in \mathbb{Q}$ and

$$\Gamma_{t_i - r_i}(\theta_{r_i}\omega) \leq \Gamma_{t_i - s_i}(\theta_{s_i}\omega) + 2^{-n}\epsilon.$$

Next choose $q_i \in \mathbb{Q} \cap [t_i, r_{i+1}[$ (if $i = n$ choose $q_n \in \mathbb{Q} \cap\,]t_n, t]$ unless $t_n = t$, in which case set $q_n = t_n = t$) such that $q_i < u_i$ where $u_i > t_i$ and $k_{u_i - s_i}\theta_{s_i}\omega \in \Omega_0$. This is possible since τ is a g.p. $\omega[0,t]$. By (A.11i), $\Gamma_{q_i - r_i}(\theta_{r_i}\omega) \leq \Gamma_{t_i - r_i}(\theta_{r_i}\omega)$. Because of (A.11iv-a)

$$k_{u_i - r_i}\theta_{r_i}\omega = \theta_{s_i - r_i}k_{u_i - s_i}\theta_{s_i}\omega \in \Omega_0\,,$$

and so $\tau' := \{(r_i, q_i);\ 1 \leq i \leq n\}$ is a *rational* g.p. $\omega[0,t]$. Now

$$\begin{aligned} J_t^{\tau'}(\omega) &= \prod_{i=1}^{n} \Gamma_{q_i - r_i}(\theta_{r_i}\omega) \leq \prod_{i=1}^{n} (\Gamma_{t_i - s_i}(\theta_{s_i}\omega) + \epsilon 2^{-n}) \\ &\leq J_t^\tau(\omega) + \epsilon \leq J_t(\omega) + 2\epsilon\,. \end{aligned}$$

Consequently $\inf\{J^\tau(\omega)\colon \tau \in r(t)\} \leq J_t(\omega)$. ∎

Let us review what we have accomplished so far. In view of (A.11ii), (A.18), and (A.24), $J_t \in \mathcal{F}_{t+}^*$, while (A.21) and (A.22) state that identically on Ω, $J_t \circ k_r = J_t$ for $0 < t < r$ and $J_{t+s} = J_t J_s \circ \theta_t$ for $s, t > 0$. Also it follows from (A.6), (A.20), and the fact that $P^\mu(\Omega_0) = 1$ for all μ [**S**, (55.9)] that the three processes $t \to J_t$, $t \to \Gamma_t$, and $t \to m_t$ are indistinguishable on $]0,\infty[$. Next we modify J in order to render it right continuous.

Clearly $t \to J_t(\omega)$ is decreasing on $]0,\infty[$ and $0 \le J_t(\omega) \le 1$ for each $\omega \in \Omega$. Consequently J_{t+} exists for $t \ge 0$ and $0 \le J_{t+} \le 1$. Define $J_0 := 1$ and note that $J_{(t+s)+}(\omega) = J_t(\omega) J_{s+}(\theta_t\omega)$ and $J_{t+s}(\omega) = J_t(\omega) J_s(\theta_t\omega)$ for *all* $s, t \ge 0$ and $\omega \in \Omega$. In what follows $\omega \in \Omega$ is fixed but suppressed in our notation. Define for $t > 0$

$$\text{(A.25)} \qquad \begin{aligned} K_t &:= J_t \cdot \prod_{0 \le s < t} (J_s / J_{s+}) \quad \text{if} \quad J_{t+} > 0\,, \\ &:= 0 \qquad\qquad\qquad\qquad\quad\;\, \text{if} \quad J_{t+} = 0\,. \end{aligned}$$

In order to justify this definition observe that $J_{s+} \le J_s$ and that if $J_{t+} > 0$, then

$$\log \prod_{0 \le s < t} (J_s / J_{s+}) \le \sum_{0 \le s < t} \Delta(s) / J_{s+}$$

where $\Delta(s) := J_s - J_{s+}$, $s \ge 0$. Because $J_{t+} > 0$, this last sum is dominated by

$$(J_{t+})^{-1} \sum_{0 \le s < t} \Delta(s) \le (J_{t+})^{-1} < \infty\,.$$

Consequently the infinite product in (A.25) is well-defined as

$$\uparrow \lim_{\epsilon \downarrow 0} \prod_{\substack{0 \le s < t \\ \Delta(s) > \epsilon}} (J_s / J_{s+}) < \infty\,.$$

It is easily checked that $t \to K_t$ is right continuous on $]0,\infty[$. We next claim that it is decreasing. Let K_t^ϵ be defined as in (A.25) except that the product is taken only over those $s \in [0,t[$ with $\Delta(s) > \epsilon$. Then $K_t^\epsilon \uparrow K_t$. Fix $0 < s < t$. If $J_{t+} = 0$, then $K_t = 0$ and so $K_s \ge K_t$. If $J_{t+} > 0$, let

$s \leq u(1) < \cdots < u(n) < t$ be the finite number of $u \in [s, t[$ with $\Delta(u) > \epsilon$. Then $J_s > 0$ and

$$K_t^\epsilon / K_s^\epsilon = \frac{J_t}{J_s} \prod_i \frac{J_{u(i)}}{J_{u(i)+}} = \frac{J_{u(1)}}{J_s} \frac{J_{u(2)}}{J_{u(1)+}} \cdots \frac{J_t}{J_{u(n)+}} \leq 1 .$$

Therefore $K_s^\epsilon \geq K_t^\epsilon$ and letting $\epsilon \downarrow 0$ it follows that $t \to K_t$ is decreasing on $]0, \infty[$. If $J_t = 0$ for all $t > 0$, then $K_t = 0$ for all $t > 0$. Suppose $J_{t_0} > 0$ for some $t_0 > 0$. Then the previous estimate shows that $\prod_{0 \leq s < t} (J_s / J_{s+}) < \infty$ if $t < t_0$, and hence this product approaches $1/J_{0+}$ as t decreases to zero. Therefore $\lim_{t \downarrow 0} K_t = 1$ in this case. Since $t \to K_t$ is decreasing and K_{0+} is either zero or one, $0 \leq K_t \leq 1$. We complete the definition of K by $K_0 := K_{0+}$. Then $t \to K_t$ is decreasing and right continuous on $[0, \infty[$ and takes values in $[0, 1]$. We next claim that K is multiplicative; that is,

$$K_{t+s} = K_t K_s(\theta_t); \quad s, t \geq 0 . \tag{A.26}$$

If $0 = J_{(t+s)+} = J_t J_{s+}(\theta_t)$, then either $K_t = 0$ or $K_s(\theta_t) = 0$ so (A.26) holds in this case. If $J_{(t+s)+} > 0$, then $J_{s+}(\theta_t) > 0$ and $J_{t+} > 0$. In particular, $K_0 = 1$, so (A.26) holds for $t = 0$, $s \geq 0$ in this case. But if $t > 0$ and $s > 0$

$$K_{t+s} = K_t J_s(\theta_t) \prod_{t \leq u < t+s} (J_u / J_{u+}) = K_t K_s(\theta_t) ,$$

because of the remarks in the third sentence of this paragraph. Letting $s \downarrow 0$ establishes (A.26) in all cases. It is clear that K inherits the properties $K_t \in \mathcal{F}_{t+}^*$ for $t \geq 0$ and $K_t \circ k_r = K_t$ if $0 \leq t < r$ from the corresponding properties of J. Finally recall that $J_t(\omega) = \Gamma_t(\omega)$ if $t > 0$ and $\omega \in \Omega_0$ by (A.20), and that $t \to \Gamma_t(\omega)$ is right continuous on $]0, \infty[$ with $[\Gamma_{0+}(\omega)]^2 = \Gamma_{0+}(\omega)$ for $\omega \in \Omega_0$. Therefore $K_t(\omega) = J_t(\omega)$ for $t > 0$ and $\omega \in \Omega_0$, and so $K = (K_t;\ t \geq 0)$ is indistinguishable from m.

Thus K has all the properties listed for J at the end of the preceding paragraph and, in addition, is right continuous.

It is necessary to make one last modification before arriving at the desired functional. The multiplicative property of K implies that $s \to K_{t-s}(\theta_s\omega)$ is increasing on $]0,t[$, and so we may define for $t > 0$ and $\omega \in \Omega$

$$\text{(A.27)} \qquad L_t(\omega) := \downarrow \lim_{s \downarrow 0} K_{t-s}(\theta_s\omega).$$

If $\omega \in \Omega_0$, $L_t(\omega) = \Gamma_t(\omega)$ for all $t > 0$ because of (A.20), (A.11iv-a), and (A.11iv-f). Therefore L and m are indistinguishable on $]0,\infty[$. In what follows $\omega \in \Omega$ is fixed, but suppressed in our notation. It is immediate from (A.27), that $L_{t-r}(\theta_r) = \downarrow \lim_{s \downarrow r} K_{t-s}(\theta_s)$ for $0 < r < t$. It now is easily checked that $r \to L_{t-r}(\theta_r)$ is right continuous and increasing on $[0,t[$. It also is evident that L inherits the properties $L_t \in \mathcal{F}^*_{t+}$, $L_t \circ k_r = L_t$ if $0 < t < r$, $t \to L_t$ is decreasing on $]0,\infty[$, and $0 \le L_t \le 1$ from K. Let $L_0 := \uparrow \lim_{t \downarrow 0} L_t$ to complete the definition of L. The proof of our final lemma is taken from the argument leading to (55.14) and (55.16) in [**S**]. We still suppress the fixed $\omega \in \Omega$ in our notation.

(A.28) **Lemma.** (i) *$t \to L_t$ is right continuous on* $[0,\infty[$.

(ii) $L_{t+s} = L_t L_s(\theta_t)$ for $s,t \ge 0$.

Proof. (i) If $0 < r < s < t < u$, then the multiplicative property of K implies

$$K_{u-r}(\theta_r)K_{t-s}(\theta_s) = K_{t-r}(\theta_r)K_{u-s}(\theta_s),$$

and letting r decrease to zero, we obtain

$$L_u K_{t-s}(\theta_s) = L_t K_{u-s}(\theta_s) \quad 0 < s < t < u.$$

Fix $t > 0$. If $L_t = 0$, then $L_u = 0$ for all $u \ge t$ since L is decreasing, and so $u \to L_u$ is right continuous on $[t,\infty[$. If

$L_t > 0$, then $K_{t-s}(\theta_s) > 0$, $0 < s < t$ since it decreases to L_t as s decreases to zero. Therefore

$$L_u = \frac{L_t K_{u-s}(\theta_s)}{K_{t-s}(\theta_s)}, \quad 0 < s < t < u,$$

and hence $u \to L_u$ is right continuous on $]t, \infty[$ because K is right continuous. Since $t > 0$ is arbitrary and $L_0 = L_{0+}$ by definition, this establishes (i).

(ii) If $0 < r < u < t$, then $K_{t-r}(\theta_r) = K_{t-u}(\theta_u) K_{u-r}(\theta_r)$, and letting r decrease to zero we obtain $L_t = K_{t-u}(\theta_u) L_u$, $0 < u < t$. If $0 \leq s < t$, let u decrease to s with $s < u < t$ to obtain $L_t = L_{t-s}(\theta_s) L_s$ because L is right continuous. Writing this for $s = 0$ and then letting t decrease to zero completes the proof of (ii). ∎

We now have proved the following theorem.

(A.29) **Theorem.** (Meyer) *Let m be an exact weak* MF *satisfying* (A.12) *and* (A.13). *Then there exists a functional $L = (L_t;\ t \geq 0)$ indistinguishable from m which satisfies for every $\omega \in \Omega$:*

(i) *$t \to L_t(\omega)$ is decreasing and right continuous on $[0, \infty[$ and takes its values in $[0, 1]$.*

(ii) *$r \to L_{t-r}(\theta_r \omega)$ is increasing and right continuous on $[0, t[$.*

(iii) *$L_t \in \mathcal{F}^*_{t+}$ and $L_t \circ k_r = L_t$ if $0 \leq t < r$.*

(iv) *$L_{t+s}(\omega) = L_t(\omega) L_s(\theta_t \omega)$ for $s, t \geq 0$.*

(A.30) **Remarks.** Recall that the first step in our construction was to replace m by an indistinguishable exact weak MF, m^e, that is $(\mathcal{F}^e_{t+})$ adapted. If m satisfies (A.12) and (A.13) and we define $\bar{m}_t = m^e_t$ if $t < \zeta$, $\bar{m}_t = m^e_{\zeta-}$ if $t \geq \zeta$ (with $\bar{m}_t([\Delta]) = 1$), then $\bar{m}$ is equivalent to m and $\bar{m}$ satisfies (A.12) and (A.13). Thus in (A.29) it is *not* necessary to assume that m is $(\mathcal{F}^e_{t+})$ adapted. It is natural to ask if L also satisfies (A.12) and (A.13). Since $[\Delta] \in \Omega_0$, $L_t([\Delta]) = 1$ is clear.

Rather than trying to trace (A.13) through the construction, it is simpler to define $L^*_t(\omega) = L_t(\omega)$ if $t < \zeta(\omega)$, $L^*_t(\omega) = L_{\zeta(\omega)-}(\omega)$ if $t \geq \zeta(\omega)$ where $L_{0-}(\omega) = 1$ and check that L^* has the properties claimed for L in (A.29). Clearly L^* is indistinguishable from m and L^* satisfies (i). The verification of (iii) and (iv) are straightforward. For (ii) fix $t > 0$. We also fix ω, but suppress it. If $t < \zeta$ and $r < t$, $\zeta(\theta_r) = \zeta - r > t - r$, so $L^*_{t-r}(\theta_r) = L_{t-r}(\theta_r)$. Hence $r \to L^*_{t-r}(\theta_r)$ is right continuous and increasing on $[0, t[$. Next suppose $\zeta \leq t$ and $\zeta \leq r < t$. Then $\theta_r\omega = [\Delta]$ and $L^*_{t-r}(\theta_r) = 1$ on the (possibly empty) interval $[\zeta, t[$. Finally suppose $r < \zeta \leq t$. Then $t - r > \zeta - r = \zeta(\theta_r)$ and so

$$g(r) := L^*_{t-r}(\theta_r) = L_{(\zeta-r)-}(\theta_r) .$$

This shows that g is increasing on $[0, \zeta[$. Let $r < s < u < \zeta$. Then

$$\begin{aligned} g(r+) &= \downarrow \lim_{s \downarrow r} g(s) = \downarrow \lim_{s \downarrow r} \downarrow \lim_{u \uparrow \zeta} L_{u-s}(\theta_s) \\ &= \downarrow \lim_{u \uparrow \zeta} L_{u-r}(\theta_r) = g(r) . \end{aligned}$$

Consequently we may suppose in (A.29) that L satisfies (A.12) and (A.13).

Recall that two functionals $\{G^i_t;\ t \geq 0\}$, $i = 1, 2$ of X are *equivalent* provided the processes $t \to G^i_t 1_{[0,\zeta[}(t)$, $i = 1, 2$ are indistinguishable. This is the proper notion of equivalence for multiplicative functionals. We shall show that if we replace "indistinguishable from" by "equivalent to" in Theorem A.29, then we may drop the assumptions (A.12) and (A.13). To this end let $m = (m_t)$ be an exact WMF of X. As explained above (A.7) we may suppose, by passing to an indistinguishable functional, that m_t is $(\mathcal{F}^e_{t+})$ adapted and that for every ω, $t \to m_t(\omega)$ is right continuous, decreasing, and takes values in $[0, 1]$. Since $P^x(\{[\Delta]\}) = 0$ for $x \in E$ we may also suppose that m satisfies (A.12) by redefining $m_t([\Delta]) = 1$ for $t \geq 0$ if necessary. Finally define $\overline{m}_t(\omega) = m_t(\omega)$ if $t < \zeta(\omega)$ and $\overline{m}_t(\omega) = m_{\zeta(\omega)}(\omega)$ if $t \geq \zeta(\omega)$. It is easy to check that $\overline{m}$ is an exact weak MF which is *equivalent* to m. Therefore we suppose that m satisfies (A.12) and

$$(A.31) \qquad m_t = m_\zeta \quad \text{if} \quad t \geq \zeta$$

in place of (A.13).

Define (we put $m_{0-}(\omega) = 1$)

$$K_t = m_t \quad \text{if} \quad t < \zeta, \qquad K_t = m_{\zeta-} \quad \text{if} \quad t \geq \zeta.$$

Then K is an exact WMF that satisfies (A.12) and (A.13). Therefore we may apply Theorem A.29 to obtain a MF, L, indistinguishable from K and having the properties listed in (A.29). Next define

$$U = (m_\zeta / m_{\zeta-}) 1_{\{\zeta < \infty\}},$$

where $0/0 = 1$. Then $0 \leq U \leq 1$, $U([\Delta]) = 1$, and $U \in \mathcal{F}^e \subset \mathcal{F}^*$. Next define

$$(A.32) \qquad V = \operatorname{ess\,lim\,sup}_{t \uparrow \zeta} U \circ \theta_t,$$

where we put $V(\omega) = 1$ if $\zeta(\omega) = 0$. It is immediate that $V \circ \theta_t = V$ if $t < \zeta$ identically in ω. Clearly $0 \leq V \leq 1$ and $V([\Delta]) = 1$. We claim that $V \in \mathcal{F}^*$. To see this let P be a probability on $(\Omega, \mathcal{F}^0)$ and define $Q := \int_0^\infty e^{-t} \theta_t(P)\, dt$. Choose $U^1, U^2 \in \mathcal{F}^0$ with $U^1 \leq U \leq U^2$ and $U^1 = U^2$ a.e. Q. Then a.s. P, the processes $t \to U^1 \circ \theta_t$ and $t \to U^2 \circ \theta_t$ agree for ℓ a.e. t. As a result $V = V^1$ a.s. P where V^1 is defined as in (A.32) using U^1. It follows readily from the proof of [**DM**, IV-38] that $V^1 \in \mathcal{F}^0$ and so $V \in \mathcal{F}^*$. Now define

$$H_t = 1 \quad \text{if} \quad t < \zeta, \qquad H_t = V \quad \text{if} \quad t \geq \zeta.$$

It is immediate that $H = (H_t)$ is a perfect MF with $H_t \in \mathcal{F}^*$ for each $t \geq 0$. We next claim that m and KH are indistinguishable. By right continuity it suffices to show that $m_t = K_t H_t$ a.s. for each t. On $\{t < \zeta\}$ this is obvious. If $t \geq \zeta$ we consider two cases. If $m_{\zeta-} = 0$, then $m_t = 0 = K_t$ for $t \geq \zeta$ and the conclusion is obvious. If $m_{\zeta-} > 0$, then $m_r > 0$ if $r < \zeta$ and it follows that $U \circ \theta_r = U$ a.s. on $\{r < \zeta\}$. Consequently using Fubini's theorem, $V = U = m_\zeta / m_{\zeta-}$ a.s. on $\{m_{\zeta-} > 0,\ \zeta < \infty\}$. Therefore $m_t = K_t H_t$ a.s. on $\{\zeta \leq t\}$. As a result m and KH, and hence m and LH, are indistinguishable. Finally we claim that $H_t \circ k_r = H_t$ if $t < r$ and, hence $H_t \in \mathcal{F}^*_{t+}$. If $t < \zeta(\omega)$, then $t < \zeta(\omega) \wedge r = \zeta(k_r \omega)$ and so $H_t(\omega) = 1 = H_t(k_r\omega)$. If $\zeta(\omega) \leq t < r$, then $k_r \omega = \omega$ and the result is obvious. Therefore $\bar{m}_t := L_t H_t$ is a MF that is indistinguishable from m and has all of the properties listed in (A.29). Moreover, assuming L satisfies (A.12) and (A.13) as we may, it is clear that $\bar{m}$ satisfies (A.31). Hence we have established the following result which I call Meyer's master perfection theorem.

(A.33) **Theorem.** *Let m be an exact weak* MF *of X. Then m is equivalent to a* MF, *m^*, having the properties listed in* (A.29). *If m satisfies* (A.13) *or* (A.31), *then m and m^* are indistinguishable, and m^* satisfies* (A.13) *or* (A.31) *if m does. We may always suppose $m^*_t([\Delta]) = 1$ for all t.*

To paraphrase Meyer [**Me74**]: One could hardly ask for more! It is, however, natural to ask why we did not begin with the n of Theorem A.6 when constructing the better versions L and m^* in (A.29) and (A.33)? After all Ω_0 defined using n satisfies $\Omega_0 = \Omega$ which would simplify the construction immensely. The point is that we would have to replace n by an indistinguishable $(\mathcal{F}^e_{t+})$ adapted n^e to begin the construction and this would destroy the "perfect" properties of n in (A.6). In general there is a trade-off between "good" measurability of an MF and having it satisfy algebraic properties without exception. This is why Meyer's theorem is so remarkable.

We now specialize these results to terminal times. Let T be an exact weak terminal time as defined in the first sentence of §4 or in [**S**, (12.1)]. Then $m_t := 1_{[0,T[}(t)$ is an exact weak MF and one may apply (A.6) to it. Let n be an exact perfect MF which is indistinguishable from m. It is shown in [**S**, (55.20)] that $n_t = 1_{[0,S[}(t)$ where S is an exact terminal time as defined in the first paragraph of §4 and $T = S$ a.s. In order to apply (A.29) or (A.33) we assume first of all that $T([\Delta]) = \infty$. (This is no restriction since $P^\mu(\{[\Delta]\}) = 0$ for each μ on E.) We introduce two conditions on T:

$$\text{(A.34)} \quad \text{(i) } T(\omega) > \zeta(\omega) \Longrightarrow T(\omega) = \infty;$$
$$\text{(ii) } T(\omega) \geq \zeta(\omega) \Longrightarrow T(\omega) = \infty.$$

If $m_t = 1_{[0,T[}$, then m satisfies (A.12) and (A.34ii) is equivalent to (A.13) while (A.34i) is equivalent to (A.31). Suppose that T satisfies (A.34ii) so that (A.13) holds for m. Let L be the functional constructed in (A.29) and satisfying (A.12) and (A.13). See (A.30). Since m_t takes only the values 0 or 1, the same is true successively of Γ_t, J_t, K_t, and L_t. See the definitions (A.7), (A.19), (A.25), and (A.27). (In connection with (A.25) note that $J_{t+} > 0$ implies $J_{t+} = 1$ and so $J_s = 1$ for $0 \leq s \leq t$.) Now define $S := \inf\{t \colon L_t = 0\}$. Since L is right continuous, decreasing, and takes only the values 0 and 1, it follows that $L_t = 1_{[0,S[}(t)$. If T only satisfies (A.34i) so that (A.31) holds for m, it is now evident that m^* defined in (A.33) takes only the values 0 or 1. Hence $m_t^* = 1_{[0,S[}(t)$ in this case also. In both cases S is an exact terminal time as defined in §4 that is also an $(\mathcal{F}_{t+}^*)$ stopping time and $S = T$ a.s. Thus we have established the following:

(A.35) **Theorem.** *Let T be an exact weak terminal time (with $T([\Delta]) = 1$) and suppose that T satisfies* (A.34i). *Then there exists an exact terminal time S satisfying* (A.34i) *such that S is an $(\mathcal{F}_{t+}^*)$ stopping time and $S = T$ a.s. If T satisfies* (A.34ii), *then so does S.*

(A.36) **Remark.** If T is a terminal time rather than a weak terminal time then (A.34i) is *automatically* satisfied even without the assumption $T([\Delta]) = \infty$. To see this suppose that $\zeta(\omega) < t < T(\omega)$, then $T(\omega) = t + T([\Delta])$ since $\theta_t \omega = [\Delta]$. In particular $T([\Delta]) > 0$. If $0 < s < T([\Delta])$, then $T([\Delta]) = s + T([\Delta])$. Consequently $T([\Delta]) = \infty$ and so $T(\omega) = \infty$. Note that this shows that $T([\Delta])$ must be either zero or infinity. However, our normalization is $T([\Delta]) = \infty$.

One may check that the additional properties of m^* and L in (A.33) and (A.29) imply that if T satisfies (A.34i) then the S in (A.35) verifies $S \circ k_t = S$ on $\{S < t\}$ while if T satisfies (A.34ii), then $S \circ k_t = \infty$ on $\{S \geq t\}$ also. However, these properties hold automatically for any $(\mathcal{F}^*_{t+})$ stopping time. (This requires in addition to (A.10) that $\mathcal{F}^0$ contains all singletons.)

(A.37) **Proposition.** *Let R be an $(\mathcal{F}^*_{t+})$ stopping time. Then $R \circ k_t = R$ on $\{R < t\}$. If, in addition, $R = \infty$ on $\{R \geq \zeta\}$, then $R \circ k_t = \infty$ on $\{R \geq t\}$.*

Proof. First note that if $\Lambda \in \mathcal{F}^*_{t+}$, then for every $s > t$ and every probability Q on $(\Omega, \mathcal{F}^0)$, $1_\Lambda = 1_\Lambda \circ k_s$ a.s. Q. But $\mathcal{F}^0$ contains all singletons $\{\omega\}$ and so $\Lambda = k_s^{-1}\Lambda$. It now follows that $\mathcal{F}^*_{t+} = \bigcap_{s>t} k_s^{-1}(\mathcal{F}^*)$. This implies that if R is an $(\mathcal{F}^*_{t+})$ stopping time and $R(\omega) \leq t$ then given $\omega' \in \Omega$ with $k_s\omega = k_s\omega'$ for some $s > t$ one has $R(\omega') \leq t$. We now suppose $R(\omega) < t$ and shall show $R(k_t\omega) = R(\omega)$. We suppress ω in our notation. Suppose $R(k_t) < r < R < t$. Because $r < t$, $k_t^{-1}\{R > r\} = \{R > r\}$, and so $R \leq R(k_t)$. On the other hand if $R < R(k_t)$ we may choose r so that $R < r < t \wedge R(k_t)$, and now $k_t^{-1}\{R < r\} = \{R < r\}$ leads to a contradiction. This establishes the first assertion in (A.37).

For the second we first show that $R \leq R \circ k_t$. Suppose for some ω, $R(k_t\omega) < R(\omega)$. Suppressing ω, the first assertion in (A.37) implies that $t \leq R$. If $t \leq R(k_t)$, then $\zeta(k_t) = t \wedge \zeta \leq t$ implies $R(k_t) \geq \zeta(k_t)$. Therefore $R(k_t) = \infty$

which contradicts $R(k_t) < R$. If $R(k_t) < t \le R$, choose r with $R(k_t) < r < t$. Then $k_t^{-1}\{R < r\} = \{R < r\}$ leads to a contradiction. Therefore $R \le R \circ k_t$. Finally suppose $t \le R(\omega)$. Then $\zeta(k_t) = t \wedge \zeta \le t$ implies $R(k_t) \ge R \ge \zeta(k_t)$, and so $R(k_t) = \infty$. This establishes the second assertion in (A.37). ■

Remark. Of course, (A.37) is closely related to Galmarino's test [**DM**, IV-101].

One may apply (A.37) to the exact terminal time S in (A.35) which is an $(\mathcal{F}^*_{t+})$ stopping time. This leads to our final version of an exact terminal time.

(A.38) **Theorem.** *Let T be an exact terminal time as defined in §4. Then T satisfies* (A.34i) *and if S is as in* (A.35), *the processes $(t + S \circ \theta_t)_{t \ge 0}$ and $(t + T \circ \theta_t)_{t \ge 0}$ are indistinguishable. If, in addition,* (A.34ii) *holds for T and T verifies $T \circ k_t = T$ on $\{T < t\}$ and $T \circ k_t = \infty$ on $\{T \ge t\}$, then the processes $(S \circ k_t)_{t \ge 0}$ and $(T \circ k_t)_{t \ge 0}$ are also indistinguishable.*

Proof. (A.36) implies that T satisfies (A.34i) and then the exactness of S and T implies the indistinguishability of $(t + S \circ \theta_t)$ and $t + T \circ \theta_t)$. If T satisfies (A.34ii), then so does S, and so by (A.37), S also satisfies $S \circ k_t = S$ on $\{S < t\}$ and $S \circ k_t = \infty$ on $\{S \ge t\}$. It follows that $t \to S \circ k_t$ and $t \to T \circ k_t$ are left continuous on $]0, \infty[$ and that $S \circ k_t = T \circ k_t$ a.s. for each $t \ge 0$. ■

If T satisfies all of the conditions in (A.38), then it is hard to imagine any situation in the theory of Markov processes where one could not replace T by S. Note that if $B \in \mathcal{E}^e$, then T_B satisfies all of hypotheses in (A.38), and if $B \in \mathcal{E}^*$ so does σ_B, the Lebesgue penetration time of B. However by (B.10), σ_B itself is an $(\mathcal{F}^*_{t+})$ stopping time. On the other hand, to the best of my knowledge, it is not known whether or not T_B is an $(\mathcal{F}^*_{t+})$ stopping time when $B \in \mathcal{E}^e$ but not in $\mathcal{E}$.

Appendix B

In this appendix we collect some facts about outer integrals and then apply them to processes. We shall not strive for the utmost generality; rather we restrict our attention to what is necessary for the applications we have in mind. Fix a measurable space $(E, \mathcal{E})$ and let μ be an s-finite measure on $(E, \mathcal{E})$. Let μ^* be the corresponding outer measure on all subsets of E. Then μ^* is a *regular* outer measure in the sense that given $A \subset E$ there exists $B \in \mathcal{E}$ with $A \subset B$ and $\mu^*(A) = \mu(B)$. Such a B is called a *measureable cover* of A. Let ℓ be Lebesgue measure on $\mathbb{R}^+$ and ℓ^* the corresponding outer measure. As usual $\ell \times \mu$ is the product measure on $\mathbb{R}^+ \times E$. Of course both ℓ^* and $(\ell \times \mu)^*$ are regular outer measures. Let $f\colon E \to [0, \infty]$ be an arbitrary positive function on E. For such an f define

$$\mathcal{U}(f) := \{(t, x)\colon 0 < t \le f(x)\} \subset \mathbb{R}^+ \times E. \tag{B.1}$$

Then the *outer integral* of f with respect to μ, $\int^* f\, d\mu$, is defined by

$$\int^* f\, d\mu = (\ell \times \mu)^*(\mathcal{U}(f)). \tag{B.2}$$

The following proposition lists a number of properties of the outer integral. They are easy consequences of the properties of outer measures. One may find proofs (in a more general situation) in **[Ra87]** for example. We write $f \le g$ a.e. for $\mu^*(\{f > g\}) = 0$, etc.

(B.3) Proposition. *Let μ be as above and let f and g with or without subscripts denote arbitrary functions from E to $[0, \infty]$. Then:*

(i) $f \le g$ *a.e.* $\Longrightarrow \int^* f\,d\mu \le \int^* g\,d\mu$.

(ii) $\int^* f\,d\mu = 0 \Longleftrightarrow f = 0$ *a.e.*

(iii) $\int^* cf\,d\mu = c\int^* f\,d\mu$ *if* $c \in \mathbb{R}^+$.

(iv) $\int^* (f+g)\,d\mu \le \int^* f\,d\mu + \int^* g\,d\mu$.

(v) $\int^* 1_A\,d\mu = \mu^*(A)$ *for* $A \subset E$.

(vi) *If* f *is* μ*-measurable, then* $\int^* f\,d\mu = \int f\,d\mu$.

(vii) $f_n \uparrow f \Longrightarrow \int^* f_n\,d\mu \uparrow \int^* f\,d\mu$.

(viii) *Given* f *there exists* $h \in p\mathcal{E}$ *with* $f \le h$ *and* $\int^* f\,d\mu = \int h\,d\mu$.

(ix) *For* $B \in \mathcal{E}$ *define* $\mu^f(B) = \int^* 1_B f\,d\mu$. *Then* μ^f *is a (countably additive) measure on* $(E, \mathcal{E})$.

Perhaps a comment on the proof of (ix) is warranted. The finite additivity comes from the definition of a set $D \subset \mathbb{R}^+ \times E$ being $(\ell \times \mu)$-measurable and the fact that $\mathbb{R}^+ \times B$ is $(\ell \times \mu)$-measurable if $B \in \mathcal{E}$. Combining this with (vii) gives the countable additivity of μ^f. It follows from (ii) that $\mu(B) = 0$ implies $\mu^f(B) = 0$ for $B \in \mathcal{E}$. Consequently if μ is σ-finite, then given f there exists $h \in p\mathcal{E}$ such that

$$\text{(B.4)} \qquad \int^* 1_B f\,d\mu = \int 1_B h\,d\mu \quad \text{for all} \quad B \in \mathcal{E}.$$

We need two more facts which we state and prove as lemmas.

(B.5) Lemma. *Let* (μ_n) *be a sequence of measures on* $(E, \mathcal{E})$ *and* $\mu = \sum \mu_n$. *Then* $\mu^* = \sum \mu_n^*$.

Proof. Given $A \subset E$,

$$\begin{aligned} \mu^*(A) &= \inf\{\mu(B) \colon A \subset B \in \mathcal{E}\} \\ &= \inf\left\{\sum \mu_n(B) \colon A \subset B \in \mathcal{E}\right\} \\ &\ge \sum \inf\{\mu_n(B) \colon A \subset B \in \mathcal{E}\} = \sum \mu_n^*(A). \end{aligned}$$

For the opposite inequality we use the fact that each μ_n^* is a regular outer measure. Thus for each n, there exists $B_n \in \mathcal{E}$ with $A \subset B_n$ and $\mu_n^*(A) = \mu_n(B_n)$. Let $B = \cap B_n \supset A$. Then $\mu_n^*(A) = \mu_n(B)$ for each n. Consequently

$$\mu^*(A) \leq \mu(B) = \sum \mu_n(B) = \sum \mu_n^*(A) .$$ ■

Return now to the situation in Proposition B.3. Since μ is s-finite, $\mu = \sum \mu_n$ where each μ_n is finite. But then $\ell \times \mu = \sum (\ell \times \mu_n)$ and consequently from (B.2) and (B.5)

$$\text{(B.6)} \qquad \int^* f \, d\mu = \sum \int^* f \, d\mu_n$$

for all $f \colon E \to [0, \infty]$.

Next let $E = \mathbb{R}^+$ and μ be an s-finite measure on $\mathbb{R}^+$. Fix $s > 0$ and let $\tau_s(t) = t - s$ map $[s, \infty[$ to $\mathbb{R}^+$. Let μ_s be the restriction of μ to $[s, \infty[$. The next lemma extends the familar "change of variable" formula to outer integrals.

(B.7) Lemma. *$\int^* f \, d\tau_s(\mu_s) = \int^* 1_{[s,\infty[} f_s \, d\mu$ where $f_s(t) = f(t-s)$ for $t \geq s$.*

Proof. By (B.3-viii) there exists $h \in p\mathcal{E}$ with $f \leq h$ such that $\int^* f \, d\tau_s(\mu_s) = \int h \, d\tau_s(\mu_s)$. But

$$\int h \, d\tau_s(\mu_s) = \int_{[s,\infty[} h_s \, d\mu \geq \int^* 1_{[s,\infty[} f_s \, d\mu ,$$

because $h_s \geq f_s$. The opposite inequality is proved similarly. ■

We now have the necessary tools to apply these ideas to processes. Let X be an arbitrary right process and let N be a HRM of X as defined in (8.2).

(B.8) Proposition. *Let $f \colon E \to [0, \infty]$ be arbitrary and define for each $D \in \mathcal{B}^+$*

$$\text{(B.9)} \qquad \gamma(\omega, D) := \int^* 1_D(t) f \circ X_t(\omega) N(\omega, dt) .$$

Then $\gamma(\omega,\cdot)$ *is a measure for each* ω *and* $\gamma(\theta_t\omega, D) = \gamma(\omega, D+t)$ *identically in* $D \in \mathcal{B}^+$, $\omega \in \Omega$, *and* $t \geq 0$. *If* $f \in p\mathcal{E}^*$, *then* γ *is a* HRM *which we denote by* $\gamma = f * N$. *For each probability* P *on* Ω, *for* P *a.e.* ω, $t \to f \circ X_t(\omega)$ *is* $N(\omega,\cdot)$ *measurable and* $\gamma(\omega, D) = \int_D f \circ X_t(\omega) N(\omega, dt)$ *as measures on* $\mathbb{R}^+$.

Proof. Since $N = \sum N_k$ where each N_k is a bounded kernel from $(\Omega, \mathcal{F}^*)$ to $(\mathbb{R}^{++}, \mathcal{B}^{++})$, it is clear that for each ω, $N(\omega,\cdot)$ is an s-finite measure on $\mathbb{R}^{++}$. Hence by (B.3), $\gamma(\omega,\cdot)$ is well-defined and is a measure (on $\mathbb{R}^{++}$) for each ω. The fact that N is homogeneous may be expressed as $N(\theta_s\omega,\cdot) = \tau_s(N_s(\omega,\cdot))$ for each $s > 0$ where $\tau_s\colon [s,\infty[\to \mathbb{R}^+$ is defined above (B.7) and $N_s(\omega,\cdot)$ is the restriction of $N(\omega,\cdot)$ to $[s,\infty[$. But using (B.7)

$$\gamma(\theta_s\omega, D) = \int^* 1_D(t) f \circ X_{t+s}(\omega) \tau_s(N_s(\omega,\cdot))(dt)$$

$$= \int^* 1_D(t-s) f \circ X_t(\omega) N(\omega, dt) = \gamma(\omega, D+s),$$

and this is obviously true for $s = 0$ (recall $\theta_0\omega = \omega$). Next suppose $f \in p\mathcal{E}^*$. Then the familiar "sandwiching" argument (see [**S**, p.11], for example) shows that the mapping $(t,\omega) \to f \circ X_t(\omega)$ is $(\mathcal{B}^+ \otimes \mathcal{F}^0)^*$ measurable. In particular if P is a probability on Ω, this mapping is $P(d\omega)N(\omega, dt) = \sum P(d\omega)N_k(\omega, dt)$ measurable where $N = \sum N_k$ as in the first line of this proof. It now follows that for P a.e. ω, $t \to f \circ X_t(\omega)$ is $N(\omega,\cdot)$ measurable and

$$\gamma(\omega,\cdot) = \int_{(\cdot)} f \circ X_t(\omega) N(\omega, dt).$$

Moreover $\omega \to \gamma(\omega, D)$ is $\overline{\mathcal{F}^0}^P$ measurable for each $D \in \mathcal{B}^+$. Since P is arbitrary, γ is a kernel from $(\Omega, \mathcal{F}^*)$ to $(\mathbb{R}^{++}, \mathcal{B}^{++})$.

It is clear from (B.9) and (B.3-ii) that $\gamma(\omega,\cdot)$ is carried by $]0,\zeta(\omega)]$ since $N(\omega,\cdot)$ is carried by this set. Finally defining

$$\gamma_k^n(\omega, D) = \int^* 1_D(t) f_n \circ X_t(\omega) N_k(\omega, dt)$$

where f_n are bounded functions increasing to f, the same arguments show that each γ_k^n is a bounded kernel from $(\Omega, \mathcal{F}^*)$ to $(\mathbb{R}^{++}, \mathcal{B}^{++})$ although not necessarily homogeneous. It follows from (B.5) and (B.3-vii) that $\gamma_n := \sum_{k=1}^{n} \gamma_k^n$ are bounded kernels increasing to γ, and so γ is a HRM. ∎

Remarks. If $f \in p\mathcal{E}^*$ and $\mathcal{F}^0$ contains all singletons $\{\omega\}$, taking $P = \epsilon_\omega$ we see that $t \to f \circ X_t(\omega)$ is $N(\omega,\cdot)$ measurable for each ω, and so one may erase the "$*$" in the definition (B.9) of $\gamma = f * N$. This is certainly the case when Ω is the canonical space of all right continuous paths in E with Δ as cemetery point. Although this is the case when (6.2) is in force, it is inconvenient to always insist on this. Sharpe in §32 of [**S**] has another approach to defining $f * N$. The present approach is somewhat more general and perhaps simpler.

(B.10) Corollary. *Given $B \in \mathcal{E}^*$,*

$$\sigma_B := \inf\{t: \ell^*(\{s \le t: X_s \in B\}) > 0\}$$

is an exact terminal time as defined in §4 and σ_B is an $(\mathcal{F}_{t+}^*)$ stopping time. For almost all ω, $s \to 1_B \circ X_s(\omega)$ is Lebesgue measurable and a.s., $\sigma_B = \inf\left\{t: \int_0^t 1_B \circ X_s\, ds > 0\right\}$. Let $B^* := \{x: P^x(\sigma_B = 0) = 1\}$. Then $\sigma_B = T_{B^*}$ a.s. One calls σ_B the Lebesgue penetration time of B.

Proof. Let

$$A_t(\omega) := \ell^*(\{s \le t: X_s(\omega) \in B\}) = \int^* 1_{[0,t]}(s) 1_B \circ X_s(\omega)\, ds\,.$$

Clearly $t \to A_t(\omega)$ is continuous and it follows from (B.8) that $A_{t+s} = A_t + A_s \circ \theta_t$ identically. Under the present hypotheses a "sandwiching" argument shows that $A_t \in \mathcal{F}_t^*$ for each $t \geq 0$. Therefore $A = (A_t)$ is a (perfect) continuous additive functional which is $(\mathcal{F}_t^*)$ adapted. It is now clear that σ_B is an exact terminal time. Since $\{\sigma_B \geq t\} = \{A_t = 0\}$, σ_B is an $(\mathcal{F}_{t+}^*)$ stopping time. The fact that $\sigma_B = T_{B^*}$ a.s. is a standard result about continuous additive functionals. See [**BG**, V-(3.6)] or (59.3) and (64.2) in [**S**]. ∎

The outer integral may be used to define quantities unambiguously in many other situations. We give one more illustration. Let $(W, \mathcal{G}^0, Q, (\sigma_t)_{t\in\mathbb{R}})$ be a flow as defined in (9.1). Let $F \in p\mathcal{G}^Q$ where $\mathcal{G}^Q$ is the Q-completion of $\mathcal{G}^0$. Define

(B.11)
$$\bar{F}(w) = \int^* F(\sigma_t w)\, dt\,.$$

A simple argument similar to the proof of (B.7) shows that for each $f\colon \mathbb{R} \to [0, \infty]$ and $s \in \mathbb{R}$

(B.12)
$$\int^* f(t+s)\, dt = \int^* f(t)\, dt\,.$$

Consequently $\bar{F}(\sigma_s w) = \bar{F}(w)$ identically in s and w. This much is valid for any $F\colon W \to [0, \infty]$. Since $F \in p\mathcal{G}^Q$ there exist $F_1, F_2 \in p\mathcal{G}^0$ with $F_1 \leq F \leq F_2$ and $F_1 = F_2$ a.e. Q. Then $F_1 \circ \sigma_t = F_2 \circ \sigma_t$ a.e. Q for each t because $\sigma_t(Q) = Q$, and the Fubini theorem now implies that $t \to F(\sigma_t w)$ is Lebesgue measurable for Q a.e. w and $\int F \circ \sigma_t\, dt$ is $\mathcal{F}^Q$ measurable. Hence so is $\bar{F}$. If $F \in \mathcal{G}^*$, then a sandwiching argument shows that $(t, w) \to F(\sigma_t w)$ is $(\mathcal{B} \otimes \mathcal{G}^0)^*$ measurable and hence $\bar{F} \in \mathcal{G}^*$.

Bibliography

PT= Probability Theory and Related Fields, continuation of ZW; SLN = Springer Lecture Notes in Math. Springer, Berlin-Heidelberg-New York; SSP = Seminar on Stochastic Processes. Birkhäuser, Boston; TAMS = Trans. Amer. Math. Soc.; ZW = Zeit. f. Wahrscheinlichkeitstheorie Verw. Geb.

Standard Reference Books

[BG] R. M. Blumenthal and R. K. Getoor, Markov Processes and Potential theory. Academic Press, New York, 1968.

[DM] C. Dellacherie et P. A. Meyer, Probabilités et Potentiel. Hermann, Paris. Ch. I–IV 1976, Ch. V–VII 1980, Ch. IX–XI 1983, Ch. XII–XVI 1987, Ch. XVII–, to appear.

[G] R. K. Getoor, Markov Processes: Ray Processes and Right Processes. SLN **440** (1975).

[S] M. J. Sharpe, General Theory of Markov Processes. Academic Press, San Diego, 1988.

General References

[B86] R. M. Blumenthal, A decomposition of excessive measures. SSP 1985 (1986) 1–8.

[BBC81] N. Boboc, G. Bucur, A. Cornea, Order and Convexity in Potential Theory: H-Cones. SLN **853** (1981).

[CL75] A. Cornea and G. Licea, Order and Potential Resolvent Families of Kernels. SLN **494** (1975).

[Dy80] E. B. Dynkin, Minimal excessive functions and measures. TAMS, **258** (1980) 217–244.

[F87a] P. J. Fitzsimmons, Homogeneous random measures and a weak order tor excessive measures of a Markov process. TAMS, **303** (1987) 431–478.

[F87b] P. J. Fitzsimmons, On two results in the potential theory of Markov processes. SSP 1986 (1987) 21–29.

[F88a] P. J. Fitzsimmons, On a connection between Kuznetsov processes and quasi-processes. SSP 1987 (1988) 123–133.

[F88b] P. J. Fitzsimmons. Penetration times and Skorokhod stopping. Sém. de Prob. XXII. SLN 1321 (1988) 166–174.

[F90] P. J. Fitzsimmons, On the equivalence of three potential principles for right Markov processes. PT, to appear.

[FG88] P. J. Fitzsimmons and R. K. Getoor, Revuz measures and time changes. Math. Zeit **199** (1988) 233–256.

[FG89] P. J. Fitzsimmons and R. K. Getoor, Some formulas for the energy functional of a Markov process. SSP 1988. (1989) 161–182.

[FM86] P. J. Fitzsimmons and B. Maisonneuve, Excessive measures and Markov processes with random birth and death. PT **82** (1986) 319–336.

[FS89] P. J. Fitzsimmons and T. S. Salisbury, Capacity and energy for multiparameter Markov processes. Ann. Inst. Henri Poincaré, to appear.

[G80] R. K. Getoor, Transience and recurrence of Markov processes. Sém. de Prob. XIV. SLN **784** (1980) 397–409.

[G84] R. K. Getoor, Capacity theory and weak duality. SSP 1983 (1984) 97–130.

[G87] R. K. Getoor, Measures that are translation invariant in one coordinate. SSP 1986 (1987) 31–34.

[GG84] R. K. Getoor and J. Glover, Riesz decompositions in Markov process theory. TAMS **285** (1984) 107–132.

[GG87] R. K. Getoor and J. Glover, Constructing Markov processes with random times of birth and death. SSP 1986 (1987) 35–69.

[GS84] R. K. Getoor and M. J. Sharpe, Naturality, standardness, and weak duality for Markov processes. ZW, **67** (1984) 1–62.

[GSt86] R. K. Getoor and J. Steffens, Capacity theory without duality. PT **73** (1986) 415–445.

[GSt87] R. K. Getoor and J. Steffens, The energy functional, balayage, and capacity. Ann. Inst. Henri Poincaré **23** (1987) 321–357.

[GSt88] R. K. Getoor and J. Steffens, More about capacity and excessive measures. SSP 1987 (1988) 135–157.

[He74] D. Heath, Skorokhod stopping via potential theory. Sém. de Prob. VIII. SLN **381** (1974) 150–154.

[H57] G. A. Hunt, Markoff processes and potentials I. Illinois J. Math. **1** (1957) 44–93.

[J87] K. Janssen, Representation of excessive measures. SSP 1986 (1987) 85–105.

[Ka78] M. Kanda, Characterization of semipolar sets for processes with stationary independent increments. ZW **42** (1978) 141–154.

[Ku74] S. E. Kuznetsov, Construction of Markov processes with random times of birth and death. Theory Probab. Appl. **18** (1974) 571–575.

[Ku84] S. E. Kuznetsov, Nonhomogeneous Markov processes. J. Soviet Math. **25** (1984) 1380–1498.

[Ma75] B. Maisonneuve, Exit Systems. Ann. Probab. **3** (1975) 399–411.

[Me68] P. A. Meyer, Processus de Markov: la Frontière de Martin. SLN **77** (1968).

[Me73] P. A. Meyer, Note sur l'interprétation des mesures d'équilibre. Sém. de Prob. VII. SLN **321** (1973) 210–216.

[Me74] P. A. Meyer, Ensembles aléatoires markoviens homogénes, II. Sém. de Prob. VIII. SLN **381** (1974) 181–211.

[Mi79] J. B. Mitro, Dual Markov processes: applications of a useful auxiliary process. ZW **48** (1979) 97–114.

[N61] J. Neveu, Lattice methods and subMarkovian processes. Proc. Fourth Berk. Symp. Math. Stat. and Prob. Vol. II, 347–391, Univ. of Calif. Press, Berkeley and Los Angeles. 1961.

[N76] J. Neveu, Sur les mesures de Palm de deux processus ponctuels stationnaires. ZW **34** (1976) 199–203.

[PS71] S. C. Port and C. J. Stone, Infinitely divisible processes and their potential theory I. Ann. Inst. Fourier (Grenoble) **21** (1971) 157–275.

[Ra87] M. M. Rao, Measure Theory and Integration. John Wiley and Sons, New York. 1987.

[Re70] D. Revuz, Mesures associées aux fonctionnelles additives de Markov II. ZW **16** (1970) 336–344.

[Ro71] Hermann Rost, The stopping distributions of a Markov process. Inventiones Math. **14** (1971) 1–16.

[Sp64] F. Spitzer, Electrostatic capacity, heat flow, and Brownian motion. ZW **3** (1964) 110–121.

[St87] J. Steffens, Some remarks on capacities. SSP 1986 (1987) 195–213.

Notation Index

AF - additive functional, 83

B^c - B complement, 3

B^r - regular points of B, 3

b_t - birthing operator, 53

$b\mathcal{C}$ - bounded elements of $\mathcal{C}$, 2

$\mathcal{B}$, $\mathcal{B}^+$, $\mathcal{B}^{++}$ - Borel sets of $\mathbb{R}$, $\mathbb{R}^+$, and $\mathbb{R}^{++}$, 5

$C(B)$, $\hat{C}(B)$ - capacity and cocapacity, 121

Con - conservative excessive measures, 8

Dex - special notation in §5, 35

Dis - dissipative excessive measures, 8

E - state space of X, 1

E_μ - Lusinian subset of E in (6.2), 51

Exc, Exc^q - excessive, q-excessive measures, 2

ess lim sup, 152

$\mathcal{E}$ - Borel σ-algebra of E, 1

$\mathcal{E}^e$ - σ-algebra generated by excessive functions, 2

$\mathcal{E}^r$ - σ-algebra of Ray Borel sets, 1

$f\mu$ or $f \cdot \mu$ - the measure $f(x)\mu(dx)$, 5

$\mathcal{F}^e$, $\mathcal{F}^e_t$, 2

$\mathcal{F}^*$, $\mathcal{F}^*_t$ - universal completions of $\mathcal{F}^0$, $\mathcal{F}^0_t$, 3

G - set of left endpoints, 136

G^*, 137, 138

$g.p.\omega[0,t]$ - good partition of $[0,t]$ for ω, 160

$\mathcal{G}$, $\mathcal{G}_t$ - σ-algebras of Y, 52

$\mathcal{G}^0$, $\mathcal{G}^0_t$, 51

$\mathcal{G}^0_{\geq t}$, $\mathcal{G}^*_{\geq t}$, 89

$\mathcal{G}^{\bar{m}}$, $\mathcal{G}^{\bar{m}}_t$, 52

$\mathcal{G}^\nu$, $\mathcal{G}^\nu_t$, $\mathcal{G}^\nu_{T+}$, 52

$\mathcal{G}^*$, $\mathcal{G}^*_t$, 52

Har - harmonic excessive measures, 47

HRM - homogeneous random measure of X, 82

of Y, 90

$\mathcal{H}^*$ - universal σ-algebra over $(H,\mathcal{H})$, 2

$\mathcal{I}$, $\mathcal{I}^m$, 66, 96

k_t - killing operators, 53, 157

L, L^q - energy functional of X, X^q, 16, 20

L_B - last exit time from B, 14, 120, 121

ℓ - Labesgue measure, 92

ℓ^* - outer Lebesgue measure, 174

M - homogeneous random subest of Ω, 136

M^* - extension of M to W, 137

MF - multiplicative functional, 151, 154

m_t, $m(s,t,\omega)$, 151

$\mathcal{M}|\mathcal{N}$, 4

N - HRM of X defined on Ω, 80

N^* extension of N to W, 89

P_t - transition semigroup of X, 1

P^q_t - transition semigroup of X^q, 2

P^q_B, 3, 23

P_T, P^q_T, 23, 22

P_N - Palm measure of N, 97

P^ξ_K, $\mathbf{P}^\xi_N$, 116

$\hat{P}^x$, $(\hat{P}^x, A)$ - exit system, 137

Pot - measure potentials, 6

Pur - purely excessive measures, 6

$p\mathcal{C}$ - positive elements of $\mathcal{C}$,1

Q_m, Q_ν - Kuznetsov measures, 52

$\mathbb{Q}$, $\mathbb{Q}^+$ - rationals, positive rationals, 5

R_T - balayage operator, 24

R^q_T - q-balayage operator, 23, 34

R^*_B - reduction on B, 29, 35

RAF - raw additive functional, 83

RM - random measure, 80

RMF - raw multiplicative functional, 154

Rλ - reduction on λ, 35

reg(T) - regular points for T, 145

r_S - balyage in terms of Kuznetsov measures, 69, 74

$\mathbb{R}$, $\mathbb{R}^+$, $\mathbb{R}^{++}$ - reals, $[0,\infty[$, $]0,\infty[$, 5

S or $S(X)$ - excessive functions, 2

S^q or $S(X^q)$ - q-excessive functions, 2

SM - stationary measure, 96

SST - stationary stopping time, 69, 74

ST - stationary time of Y, 74
  of a flow, 97

STT - stationary terminal time, 74

T_B - hitting time of B by X, 3, 71
  by Y, 74

$t_n \downarrow\downarrow t$, 5

U - potential kernel of X, 2

U^q - resolvent of X or (P_t), 1

U_N, U_N^q, U_A^q - potential kernal of N or A, 82, 83

u_A^q, u_N^q - q-potential of A or N, 83

$\hat{V}$, 145

W - sample space of Y, 51

w - generic point of W, 51

$W(h)$, $W_t(h)$, 57

WRMF, weak raw multiplicative functional, 151

WSM, weak stationary measure, 111, 112

X, X_t - realization of P_t under (6.2), 53

Y, Y_t - coordinate process on W, 51

Y_t^* - modification of Y_t at α, 57

α - birth time of Y, 51

β - death time of Y, 51

$\Gamma(B)$, $\Gamma_m(B)$ - capacity of B, 120

$\Gamma^q(B)$, $\Gamma_m^q(B)$ - q-capacity of B, 130

γ_B, γ_B^m - capacitary measure of B, 126

Δ - cemetery point, 1

$[\Delta]$ - dead path 1, 51

ϵ_x - point mass at x, 49

ζ - life time of X, 1

θ_t - shift operator for X, 1
  truncated shift on W, 53

θ_t^*, 57

λ_B - restriction of the measure λ to B, 10

λ_B - last exit from B by Y, 120, 121

$\{\lambda \leq \mu\}$, 36

$(\lambda - \mu)_+$, 36

$\mu(f)$, $\langle\mu, f\rangle$, $\mu f - \int f\, d\mu$, 5

ν_N^ξ, ${}^q\nu_N^\xi$, ${}^q\nu_A^\xi$ - Revuz measure, 85

ξ_i, ξ_p - invariant, purely excessive part of ξ, 7

ξ_c, ξ_d - conservative, dissipative part of ξ, 8

π_B, $\hat{\pi}_B$ - capacitary, cocapacitary measure, 124

σ_t - shift on W, 51

σ_B - Lebesgue penetration time of B by X, 29, 178
  by Y, 74

τ_B - hitting time of B by Y, 73,120

τ_B^*, 128

$\varphi \otimes f$ - φ tensor f, 93

φ^+ - right hand derivative of φ, 131

Ω - sample space of X, 1
  under (6.2), 53

Ω_0, 152

$\prec$ - strong order, 37

$\curlyvee$, $\curlywedge$ - sup, inf in strong order, 38

$\int^*$ - outer integral, 174

Subject Index

additive functional, 83
raw, 83
Alternating of order n, 123
of order infinity, 123
balayage, 34
Borel right process, 51
capacity, 120, 121
q-, 130
co-, 121
capacitary measure, 124, 126
m-, 126
co-, 125
cocapacity, 121
conservative
excessive measure, 8
measure of a flow, 113
cotransient set, 122
m-, 122
strongly, 128
decomposition
conservative-dissipative
for excessive measures, 8
for flows, 114
harmonic-potential, 47
invariant-purely excessive, 7
dissipative
excessive measure, 8
measure of a flow, 113
set, 9
energy functional, 16
of X^q, 20
entrance law, 43, 50
at u, 50
entrance rule, 50
equilibrium set, 126
m-, 126
equivalence
of stationary measures, 96
of functionals, 168
exact
multiplicative functional, 154
stationary terminal time, 77
regularization of, 77
terminal time, 22
regularization of, 22
excessive
function, 2
measure, 2
purely, 6
random variable, 68
exit system, 137
fine topology, 3
-ly open, 3
flow, 95
good partition for ω of $[0,t]$, 160
harmonic excessive measure, 47
hitting time of B
by X, 3, 71
by Y, 74
homogeneous
random measure
of X, 82
of Y, 90
random set, 136
Hunt's balayage of excessive measures, 34
increasing process, 81
invariant σ-algebra
on W, 66
of a flow, 96
Kuznetsov measure, 52
last exit time, 121
Lebesgue penetration time
by X, 29, 178
by Y, 74
measure
capacitary, 124, 126

cocapacitary, 125
homogeneous random
of X, 82
of Y, 90
Palm, 97, 99, 116
random, 80
Revuz
of a HRM of X, 85
of a HRM of Y, 93
of a RAF, 85
s-finite, 83
stationary, 96
weakly-, 111
σ-integrable-, 98
translation invariant along $\mathbf{R}$, 92
m-cotransient, 122
m-polar, 122
m-semipolar, 146
m-transient, 122
natural order, 37
outer integral, 174
outer measure, 174
regular-, 174
Palm measure
of a SM, 97
of a ST, 99
of a HRM, 116
potential measure, 6
proper (potential) kernel, 2
purely excessive measure, 6
q-capacity, 130
q-potential
of N, 83
of A, 83
operator
of N, 82
of A, 83
random measure, 80
equality of, 82
homogeneous-
of X, 82
of Y, 90
raw additive functional, 83
reduction of an excessive measure, 29
respect of Q, 105
Revuz measure
of a HRM of X, 85
of a HRM of Y, 93
of a RAF, 85
s-finite measure, 83
simple order, 37
solid cone, 12, 49
specific order, 37
Spitzer's formula, 143
stationary measure, 96
stationary
stopping time, 63, 74
terminal time, 63, 74
exact, 77
time, 63, 74
of a flow, 97
stopping times for Y, 52
strong order, 37
strong subadditivity, 123
strongly cotransient, 128
supermedian function, 2
regularization of, 2
terminal time, 22, 71, 171
exact, 22, 171
regularization of, 22
weak, 22, 171
totally thin, 147
transient
process, 2
set, 122
translation invariant along $\mathbf{R}$, 92
weak
raw multiplicative functional, 151
exact, 154
stationary measure, 111
terminal time, 22, 171
Weil's energy formula, 119, 120
ξ-integrable HRM, 86

σ-integrable HRM, 86
σ-integrable SM, 98
σ-Q-polar, 105

Probability and Its Applications

Editors

Probability and Its Applications includes all aspects of probability theory and stochastic processes, as well as their connections with and applications to other areas such as mathematical statistics and statistical physics. The series will publish research-level monographs and advanced graduate to Progress in Probability, a context for conference proceedings, seminars, and workshops.

We encourage preparation of manuscripts in LaTeX or AMS TeX for delivery in camera-ready copy, which leads to rapid publication, or in electronic form for interfacing with laser printers or typesetters.

Proposals should be sent directly to the editors or to: Birkhäuser Boston, 675 Massachusetts Avenue, Suite 601, Cambridge, MA 02139, U.S.A.